### You've Been Freq'ed!!

Copyright © 2018 by Ronnie D. Johnson

The reproduction, transmission or utilization of this work in whole or in part in any form by electronic, mechanical or other means, now known or hereafter invented, including xerography, photocopying and recording, or in any information storage or retrieval system is forbidden without written permission, excepting brief quotes used in reviews.

Extravagant Publications, LLC is registered in the United States Patent and Trademark Office.

All **Extravagant Publications, LLC.** Titles, Imprints and Distributed Lines are available at special quantity discounts for bulk purchases for sales promotions, premiums, fund-raising and educational or institutional use.

Imprint: *Extravagant Publications, LLC.*
www.ExtravagantPublications,LLC.com
ISBN 978-0-578-47700-8

## Acknowledgments:

I like to thank all of my families for being there for me. All the friends, who have waited so long for the book to be in print. I like to thank Mr. Tony Peay and Mr. Rodney Kimble for sticking by me once again. And a special thanks to Tammy Rowe and Nurses with Purposes for sharing their adventures abroad to help invigorate me back into writing. Thanks to Johnson Controls team for allowing me into their working family.

Without Earth's foundation none of this is possible, so now I will use what I have learned as a gift to educate others.

**Lamonte Jordan** – Graphics

GruvyGraffics.com

# Ronnie D. Johnson Bio

Nickname: RonnRamm

At the age of fifty, Ronnie D. Johnson has had a many titles added to his name - not all of them good, but the experiences outweigh the persona. He started his career out as a teenager in the US Navy aboard the super carrier John F. Kennedy. After a four year career onboard the ship, he ended his enlistment and return to his home in Memphis, Tn. He started to pursue a career in music while working for ADT Security Company. While working for ADT , Ronnie found a passion for computers and started to develop his skills using technology as his focal point. After leaving ADT, Ronnie acquired a position with the City of Memphis Police Department.  Ronnie used his amazing skills of technology and his uncanny knack to finish what he started to propel himself to the upper level of the department that he was assigned to. After leaving the police dept.

Ronnie put together a fictional persona that he made up loosely based on himself. "Frequency Hunter" would be Ronnie's alternate ego with an unlimited potential of growth and depth. Frequency Hunter is a mix between Microsoft (Bill Gates), who he greatly respects with a touch of MacGyver and a Navy Seal. As Ronnie life progressed his quest to bring "Frequency Hunter" to print began. After starting a successful business, working a full time job, plus dealing with the day to day antics of his two gorgeous children (Eboni and Ronnie Jr). Ronnie D. Johnson has brought to print a multitude of stories that will entice your passion to read.

## You've Been Freq'ed

### By

### Ronnie D. Johnson

# Table of Contents

1. Assignment
2. New Business
3. Back 2 Business
4. Fireman
5. Dump Truck
6. Thunderhead Hawkins
7. Bad Man
8. Switch On
9. N Da Game
10. White, Black , and Red all Over
11. She Didn't Get It
12. Quarter, Nickel, and Dime
13. Teacher ,Teacher
14. Car Wreck
15. Geeks & God
16. Junk N Trunk
17. Correbous
18. Done Deal

# F.R.E.A.K.S.

Federal Response Enforcement Against Kid Sex

**Ronnie D. Johnson**

# You've Been Freq'ed!!"

# 1.) <u>ASSIGNMENT</u>

Frequency Hunter sat in his chair, just thinking back on recent events in his life. He had been wanted by the Fed's for hacking and cracking computers. Which resulted in him being arrested. He then spent time in jail and then was released. Upon his released he found out that he had been doubled crossed by a close friend. He then found himself on the run from the Feds because he cracked the computer of a major pedophile ring that was under investigation. Frequency laughed when he found out that the Judge that sent him to jail was into sex with young people. He loved the look on Judge Hill's face when he found out that Frequency turned over all his encrypted files to the Feds. Frequency still had some unfinished business with the rest of the Judges organization. Frequency was very upset when he found out that GayMack, Ms. Bobbi and Madame had escaped the Feds. Frequency released the thought when Agent Layrock and Masters came to him with a new case that needed the agencies' handy work.

Frequency and his team of specialist had joined forces with the Feds to make up a team called F.R.E.A.K.S. Which was short for Federal Response Enforcement Against Kid Sex. Their unique task was to breakdown and arrest any and all person, whom committed heinous sex act against children. Some of the cases were easily cracked and then there were some that brought out the best in Frequency and his team. Frequency sat inside LongHaul and listened to the description of the scene of evidence gathered by another team of operatives. Agent Layrock said this case would be a hard one since the person, who was assaulted was well connected on the streets. After doing more investigating the agents found out that this had been going on for time and all the attacks were on young men between the ages of 13 and 25. Frequency asked Mack Master how many and Mack Master had CMAX pull the info from the file. There were over 25 reported cases. Frequency told CMAX to map out the locations by time date and location. Frequency was surprised to see "How wide spread it was"? Agent Layrock then said the disturbing fact is that this is only the reported cases. A lot of the men aren't coming forth since they did not want it to be known that they had been raped. Frequency immediately contacted the rest of the crew and told them that they had a new assignment.

EoW was just getting to the computer shop and said that he would not be able to meet for about an hour. Don1, Webb, and BDB all agreed. So, Frequency told all of them to meet exactly in one hour at the new facility. All the fellows agreed and signed off. Frequency immediately began doing research to find out anything about the case that might have been overlooked. Frequency's research found that the attacker left no DNA of any kind at the scene or on the victim. Frequency dug deeper and he found out that the victims were drugged and then dropped elsewhere. Frequency needed to go to the scene where the victim was attacked, and he needed to talk to the victims. Frequency stopped for a second to laugh at himself. Here he was not long ago running from the Feds, but now he was thinking like one. His thought was interrupted when the overhead door to the facility opened and BDB's truck pulled in. BDB got out of the truck and immediately walked toward LongHaul. He stepped inside and shook hands and hugged Frequency. Frequency told BDB that he was early and BDB said he wanted to keep it that way.


   Next came Webb and EoW. They both had what was left of lunch and stepped inside LongHaul.Don1 was last and all the crew knew why he was last. EoW cracked on him as soon as he stepped foot inside of LongHaul. EoW said that the shade of lipstick didn't go at all with his outfit. Don1 went by a mirror and cleaned the lipstick off his lips. Don1 sat down and said that now he was here could they start the meeting. With that Frequency called in Agent Layrock and Master. The two men came in and told the crew what they had and what they needed to shut this case down. All the fellows had ideals and Frequency made sure that CMAX ran the ideals for logic. As the ideals came across the screen Frequency and the crew setup parameters to work the case by. Don1 told EoW that it looked as if he was going back into an area that he did not like. EoW swallowed hard and said that if he did not have to do business with the owners, he could make it end and out with no problem. EoW then asked Agent Layrock could he go back inside with agent Jones since now they were on the same team. Agent Layrock said that he was afraid not, due to agent Jones is out on another assignment working with her husband. EoW made a face as is he bit a sour lemon and the crew just laughed. Frequency then told BDB that he would be working closely with him on this case.

Frequency instructed Webb that he would be needing some high-profile street rides to pull off a major ploy in the hood. Webb asked Frequency did he need rides that were built for speed or strength. Frequency had CMAX pull the file on all the victims, who had come forward about their assaults. Frequency pulled one. Mr. Frank Gray. The crew looked at one another because this young man had drawn a blank. Frequency then showed, who this young man was related to. It was a big-time street hustler named C4. When Frequency showed C4's picture his whole crew knew what time it was. They knew that C4 would have his people out on the street banging for information. That meant that innocent people would be hurt once he started looking for, who did this to one of his family members. Agent Layrock stated that he could send out a car to pick up C4, but Frequency was immediately against that. He told agent Layrock that the last thing that they needed was for him or any agents to go around picking up anybody. Frequency then told Agent Mack Master that he would have to play his old role of probation officer, because most of the street people knew him just for that. Agent Layrock laughed and said see what happens when you are good at your job. It just pulls you right back in. Frequency then gave Agent Layrock his assignment.

He told agent Layrock that he was going under cover as a street racer named LROCK. Agent Layrock said that he did not know a thing about racing cars. Frequency said that is precisely what he wanted. He needed somebody that was green to the game and he fit the bill perfectly. We will set you up with everything that you need and plus you will always have CMAX with you. Agent Layrock smiled and asked will he be able to dress the part and Frequency said of course. Frequency then instructed Webb to put together a Lexus street racer for the newest street racer LROCK. The fellows all laughed when Agent Layrock borrowed some shades from EoW and folded his arms. Frequency then told Agent Layrock to go out spend money on clothes and jewelry at these stores in the hood.

He told agent Layrock to use cash and if confronted by anyone tell them to take it to the streets. More than likely Frequency said you will find out where the races are and then that is when you will bring in your crew to wreck shop. Agent Layrock asked what that meant, and Webb stated that the real street racers come in and win what you lose. Agent Layrock said to Webb "who said that I am going to lose". Webb stated if you don't lose, I will hook your ride up Freak style, but if you lose you will drive any vehicle of my choice for the duration of the case. Agent Layrock agreed and then he left LongHaul to change out of his usual black suit.

  Webb immediately had CMAX search the police impound lot for a Lexus SC coupe. It found several that were totaled, but it had two that were salvageable. Webb gave Mack Master the two lot numbers and told him that he would need his expertise to get them from the impound lot. Mack Master said no problem and he headed straight for the lot. Webb then asked BDB would he stay and help him with building this car because he would surely need the muscle. BDB said sure and then asked CMAX for all the information on How to hotrod one of these cars. CMAX brought up hundreds of diagrams to make this Lexus one of a kind. BDB then laughed and said are we building agent Layrock a way to kill himself. Webb laughed and said I hope Frequency has something in mind to help the old boy because he is going to need it. Don1 was already ordering communication equipment for the car and he made sure that the car had 21 different points for GPS to track from. EoW was designing an interface for CMAX to damn near drive the car like something off the show "Knight Rider". The crew was well into the research when Mack Master showed up with both cars. Webb looked at both and decided to build two cars exactly alike. One for the quarter mile and the other for drifting. That way no matter what LROCK got himself into he had the equipment to get himself out of.

  Frequency entire crew was at work preparing for the assignment. He had a constant monitor on all the local police radios trying to hear anything pertaining to the" FREAKS" latest assignment. The dispatcher came over the radio and stated that a body of a fourteen-year-old boy was found behind a nearby construction site. Frequency told agent Mack Master to take him to the scene immediately. Mack Master went and jumped into his old Monte Carlo and it would not start. Frequency pointed to Webb to take care of this as he started to laugh. Frequency then gave Mack

Master the keys to agent LayRock Mercury Marauder. Mack Master got into the car and started it without any problems. Frequency got into the car and Mack Master headed straight for the scene of the crime. Frequency was all shook up by the time that they had arrived. Mack Master was taken aback by the power of the Marauder. Frequency just laughed and said that he could drive it until agent Layrock was through with his undercover work. Frequency then told Mack Master to wear a special pair of shades that had a link straight into CMAX.

  Mack Master asked Frequency was he coming, and he said no. Frequency informed Mack Master that he needed to stay low key since he was a part of LROCK crew. Mack Master nodded that he understood and then he went on to the crime scene.

Police forensic were all over the place and Mack Master followed each one of them. Nobody paid much attention to Mack Master until he got closed to the body. One Officer Poole immediately told Mack Master to stand back because the person was alive. Mack Master pulled out his badge and the officer backed off. The victim lay on the ground motionless. His pants were around his ankles and there was blood present at the victim's rectum. The young man seemed to be about 15 years old and had a tattoo on his upper forearm. Over Mack Master's shoulder he heard someone say that this young man is affiliated with one of the local gangs. Mack Master turned to see the face of detective Morris. She had been assigned to serial assault case. Det. Morris stated that she had been following this maniac for months. It was the same in all cases and she was willing to bet that there was no evidence to be found. Det. Morris and the police forensic lab had collected all that they could and then they let the paramedics take care of the person. Det. Morris told Mack Master that there would be retaliation for this attack. Mack Master asked Det. Morris, who would they attack and why? Det. Morris said they would attack the gay community because they are the likely suspects. Mack Master thought for a second and then shook his head. He knew she was right, and he hoped Frequency heard her. Mack Master waited until the victim was loaded into the ambulance and then he went back to the car where Frequency was. Frequency immediately told Mack Master that he should go to the hospital and see who shows up to see about the boy. Mack Master agreed and headed straight to the hospital where the victim was taken. Frequency went into the waiting room and picked up a newspaper and pretended like he was awaiting treatment. Mack Master went up into the emergency room area to see if he could gather some more information. Frequency watched as a gang of people stormed into the hospital and asked about the little guy who had been assaulted. Frequency overheard the name Mario Perez and then he heard the nurse say that he was just coming out of the ER. The leader was heavily tattooed, and he had a mouth full of gold teeth. One of the other men called him PPK and Frequency had CMAX run the name through all the crime computer's databases.

The name "PPK" came up as Peter Perez King of the Latin Machine Gang. It was the largest gang in the region. Frequency saw that the LMG was one of the most powerful of all the gangs and they controlled majority of the street traffic period. Here now in front of Frequency was LMG1 and Frequency had a full head to toe shot of him.

PPK was out of his environment, but he did not care. He had his best soldiers with him if anything should go down. PPK knew he could not shake down the nurse in the hospital, but he knew she would not stop him from seeing his little brother. PPK asked the nurse in a calm voice could he see his little brother. The nurse knew she could not stop him and so she told him that he was on the 16$^{th}$ floor room 9. PPK told the nurse thank you and then he headed for the room with his entire crew in tow. They bombarded the elevators and headed straight to the 16$^{th}$ floor. Before long majority of the crew was on the 16$^{th}$ floor and hospital security was overwhelmed by the sure mass. Det. Morris showed up and was met face to face by PPK. PPK had some question for her to answer and she was going to tell him by hook or crook.

Det. Morris introduced herself to PPK and told him everything that she knew. PPK immediately thought that it was a rival gang, but Det. Morris immediately squashed that when she told him that every neighborhood in the city had been hit by this serial rapist. PPK asked Det. Morris why haven't they done something about it? Det. Morris told PPK that majority of the victims would not come forward because they are men, who do not want the public to know that they have been raped. Det. Morris then asked PPK would he come forward if some man had left him somewhere with his pants around his ankles and his rectum bleeding. None of the gang members said a word as PPK held his temper and did not say a word. He then stepped close to Det. Morris and said something for her ears only. Det. Morris held her composure and then said if she found out anything he would be contacted. PPK then made a statement out loud so that all his members could hear him. He said that he was putting up a 100K to anyone who finds out who did this to his little brother. PPK then went in to see his brother. The doctors had to put him on some medicine to counteract the drug he was given during his attack. The doctor told PPK that his brother must have been one hell of a fighter. PPK asked the doctor why? The doctor said that Mario was one of the 1$^{st}$ victims that he had to treat that had been drugged.

PPK smiled to himself knowing that his little brother went down fighting. He knew that now he had to get this idiot who put his hands on his little brother.

PPK told the doctor to do whatever he needed to do to get his brother well. PPK said that he had all the bills covered and it would be paid as soon as his brother was released to his care. PPK then left the room. He told two of his members to stay with his brother no matter what. The two men nodded ok and then PPK and the rest of the gang left the hospital.

Mack Master watched in awe as the LMG members got into their cars and left single file. Mack Master saw low riders of all kinds and he even saw a Monte Carlo like his, except it was in much better shape. Mack Master looked at Frequency and Frequency immediately knew what he wanted. Frequency told Mack Master to hook up with Webb and he would tell you if he can? Mack Master smiled as he and Frequency headed back to the facility to see the progress on the car and the transformation of LROCK.

The crew were surprised to see Agent Layrock in street gear. He looked out of place, but with a few minutes of gaining his composure, He was sure to fit in. Frequency asked LROCK what he was he rocking, and he said that he was Sean John from head to toe. Frequency nodded his approval. LROCK had did his homework and had the slang of the streets down packed. LROCK then told Frequency all about the car he was driving. LROCK did not miss a pronunciation of the refinements made on the car. LROCK then got into the car and said that it was an extension of his hands. Webb interrupted the two men and told them that all that talking ain't crap without backup. LROCK jumped into the car and took it out into the public. Lucky for them nobody was out to interfere, and LROCK began to show off his driving skill. 1ˢᵗ he did some high power figured eights and then he took off down the road just to see "what she would do."

EoW had shut down the shop and headed over to Don1's place. They met up outside and decided to go to a local restaurant for dinner. They decided to trail one another to the restaurant just in case one of them needed to partake in the night's festivities. EoW and Don1 sat down at a Chili's and order an appetizer and drinks. They were going over all the details of LROCK's new ride when they got the call from Webb. Webb told them that all was a go and that LROCK was in the streets. Webb also told them that Frequency had left in a cab and he had not spoken to him. Don1 asked Webb where was BDB. Webb stated "Good Question, but he did not have an answer. Don1 told Webb to shut down for the night and he would find out what was going on. Don1 patched into CMAX and had him use the GPS in Frequency's phone to track him. The GPS showed that Frequency was not too far from the restaurant.

Don1 told EoW that Frequency had taken a cab and left the station. He then told him that he was up the street from the restaurant. Don1 told EoW to place his order for him and he would go get Frequency.

Frequency paid the cab driver and got out of the car at an independent car lot. He went onto the lot, but they had closed for business. Frequency walked down the inventory of cars until he walked up on what he was looking for. It was a Pontiac Trans AM like his old car except this one did not have T-tops. Frequency kicked the tires and walked around the car looking at the wear tear of the exterior. The Night watchman came out to tell Frequency that the business was closed. Frequency asked him about the car and the night watchman was brutally honest. He said that he would not give that car to his mortal enemy. The night watchman told Frequency that the car was used in a robbery several months ago and was shot to hell. The dealership body shop assistant learned how to do bondo on this car. And now the owner is trying to find a sucker to drop it on. Frequency smiled at the night watchman and thanked him for the information. The night watchman said that he did not need any thanks. He just felt like nobody should be taken like that. Frequency asked the night watchman his name and he told him Mr. Eddie Wright, but they call me Mr. Eddie.

Frequency pulled a $100-dollar bill out of his pocket and gave it to Mr. Eddie. He told Mr. Eddie that he would not forget him. Mr. Eddie took the money and said thanks. He then told Frequency that he thought his ride had pulled up.

Frequency turned to see Don1's ride pulls up at the front of the lot. Frequency waved to him and told Mr. Eddie that he would see him again. Mr. Eddie said sure and walked back to his guard station. Frequency walked over and got into the car with Don1. Don1 immediately started questioning him. Frequency answered Don1 with questions with one big answer. He told Don1 that he missed his TA. Don1 laughed hard and said I know you were not going to buy that piece of shit from "ALL GREAT AUTO SALES". Frequency laughed and said damn I forgot about that. All good auto sales were notorious for selling income tax cars. The cars basically lasted from one tax to the other if you were lucky. Don1 headed back to the restaurant where EoW was sitting waiting. EoW had taken the liberty of ordering Frequency a strawberry martini. He ordered Don1 a top shelf martini and got himself a blue Laguna martini. The three men toasted to future endeavors and then Don1 started in on Frequency. EoW added his two cents in and then they ordered their food. The three of them finished up their meals and then EoW told Frequency that he was going by his old house. Frequency asked him what for?


And EoW told him that he was curious about something. The three men headed for Frequency's old house. Frequency rode with Don1 and EoW drove his own vehicle. Don1 and EoW drove their vehicles hard and the cars performed immaculately. It was not long before they arrived at the house and got a chance to see the damage for themselves. Don1 was in shock that the house was burned so thoroughly. EoW got out of his car and grabbed a flashlight. He went to the back of the house and half of the supports were still holding. EoW pushed a few burned 4 x6s out of the way and there was what he was looking for. EoW called out to Frequency and Don1 and told them to come here. Don1 and Frequency went to the area where EoW was standing and then they asked him what was he looking for? EoW asked Frequency was this the place where he parked his vehicles. Frequency said yes, but why you ask? EoW said when you had crashed your bike fooling with Fire and Ice the police found the crashed frame and took it to the police impound lot. Frequency said yes so what is your point. EoW then asked Frequency where the burned frames of your two cars is. Frequency thought for a second and then he realized that he was missing two cars. He started to smile when he thought about the ideal that the TA was not damaged by the fire, but on the other hand where was it.

Frequency activated the voice recognition program on his cellular phone and then he linked in to CMAX. He had CMAX run traces on his two missing vehicles.

CMAX came back and told Frequency that there was no trace of his vehicles. Frequency told CMAX to continue to scan until the vehicle showed up. CMAX stated "Acknowledge" and then Frequency signed off. EoW then walked around looking for even more evidence. Don1 was shocked that the GPS system that he had set up in Frequency's cars was not working. Don1 took that information as a challenge and he was not one to back down from one. Don1 told Frequency that he had put the best GPS system in those cars and they had to be picked up by Martians not to be able to pick up a signal. Frequency thought for a moment and then told the fellows to head for EoW's place. EoW said cool and then they got back to their vehicles and headed for EoW's place. They got to EoW's condo and the men went inside and immediately started discussing the missing vehicles. Frequency brought CMAX online and asked him to track all the team's vehicles. CMAX stated that BDB was at the movie theater out east, Webb was parked at the shop, Mack Master was at his home, Don1 and EoW was at the same location as they were. And then he said LROCK is moving at a high rate of speed on the interstate near the airport. Suddenly CMAX stated that he had lost communications with LROCK. Don1 asked CMAX to use the highest level and then CMAX told Don1 that he was. CMAX then said that he had LROCK back on the interstate. He had just pulled in for gas on the other side of the Airport tunnel. Don1 smiled and shouted that he now knew what was wrong. Frequency asked Don1 what was it? And Don1 stated that the vehicles are underground or under a heavy leaded structure. Frequency said that explained why they could not find the cars, but where in the world could they be? EoW said we don't have to find them, they will tell us where they are. They cannot keep them under whatever forever. True said Don1 all we must do is wait. Frequency said that he was really curios about who would touch his cars. Those were his prize possession except for his love, Doc.

Frequency let the thought falter off and then he told the men that he wanted to go out and see "What the night had to offer." Don1 immediately told EoW that he was driving, and that Frequency was in the back. EoW laughed and said Don1 was pissed because his car outperformed the Caddy. Don1 laughed and said we will see about that. So, they got into the Caddy and headed for the airport tunnel. Don1 pushed the Caddy hard and it performed genuinely.

EoW saw the speedometer top 100 mph as they went into the airport tunnel. Suddenly Frequency told Don1 to stop and activate ONSTAR. Don1 slammed on the brakes and the big car stopped on a dime. The ONSTAR system activated but was unable to connect. Don1 tried all his tracking systems and was unable to connect. Frequency tried to connect to CMAX and was not able to connect also. Frequency told both guys that they have found their problem and now needed a solution. Don1 and Eow both made a note in their PDAs to figure out a dead spot problem in the airport tunnel. Frequency told Don1 to drive on and he got the Caddy moving again. As soon as they got out of the tunnel the ONSTAR system connected. Don1 placed a dead spot report to the representative and she stated that area was a dead spot that had been reported before and they were working on getting it resolved. Don1 disengaged the ONSTAR system and Frequency told Don1 to stay on top of it. EoW then told Don1 to head over to a local street so that he could see "who was racing for cheese". Don1 went down to the area on the other side of the airport tunnel on a street called "Swinea". The cars and motorcycles were lined up down the side as different race club made the challenges then raced down the street. EoW saw that one of the local clubs had the police scanner so that meant that the police were paid off for a few hours.

   Don1, EoW, and Frequency got out of the car and watched the races from the street side. There were some nice-looking women walking up and down the street and the fellas were trying to put there mack down. Frequency overheard some fellas from one car club talking about a new dude on the scene that had knocked off two heavy hitters today. Frequency got EoW attention and had him follow them to find out more. EoW went up the street and then called Don1 to tell him that the new dude name was LROCK. Frequency told Don1 to tell EoW to head back so that they could leave. As the three got into the car they saw LROCK pass by beating another car by 4 car lengths. All three of them started to laugh as EoW said these fellas don't know that "they've been freq'ed"!!!

## 2.) New Business

Frequency awoke the next morning at EoW's place. He and Don1 both crashed at EoW's condo since he had more room than he needed. Business had started to boom for them ever since they had been on TV for knocking down Judge Warren Hill. They had both brought new cribs and they both did not have a stitch of grass to cut. EoW's computer shop stayed busy from the time it opened until the time it closed. EoW had to expand the size of his shop by 2000 square feet. EoW now had 20 full time employees and he was known throughout the city. Don1 was also enjoying the publicity of the trial. He had a total of 4 full service cellular shops that provided services for all cellular networks. Don1 had become a major player in the cellular game in the city. He provided all the street people cellular service and they made sure he could move throughout the city without a problem. Frequency was proud of his two friends and he enjoyed watching them embrace their success. Frequency thought about how BDB landscape business had doubled two-fold. Webb had so many customers that he barely had time to hang out.

Frequency then thought about how he did not have a business of his own. He owned a percentage in all his friends' businesses, but he did not have one of his own. Frequency decided to get himself some business. He got up from EoW's couch and took a shower. He put back on the clothes that he had on from the day before and then waited for Don1. Don1 did the same and the three men went out for breakfast. After breakfast Don1 took EoW back to his condo and told him he would call him later. EoW said that he was going to the shop and then he would be in touch later. Don1 headed to his zero-lot home. He too chose a home that did not have any grass to cut. Other than that, he had a beautiful home. Don1 had his house laid like something out of a magazine. He made sure that his clothes stayed laundered and nothing was out of place. Don1 left Frequency the key to the Caddy and said that he was going in to get changed. Frequency knew that meant he had two hours to go and get a change of clothes. Frequency headed for the station where LongHaul was parked. Frequency headed for the station but ended up taking a detour due to road construction. Frequency ended up on an older side of town that he had not been on in years. He saw Billy's clothing store and decided to stop.

    Frequency went inside and immediately saw a linen set that he liked. As soon as Billy saw Frequency he came over and gave him a firm handshake and a hug. Frequency smiled and told Billy that he did not know he was still in business. Billy said that the hood keeps it good in his pockets. Frequency told Billy to set him from head to toe. Billy put together the linen set with shoes and threw in a pair sock for old times' sake. Frequency went into the dressing room and changed clothes. He stepped out from the dressing room as clean as a whistle. Frequency threw his other clothes in the trash and paid Billy for the clothes and gave him a nice fat bonus. Frequency promised Billy that he would do more business with him and walked out of the store. Frequency looked at his watch and he still had an hour and fifteen minutes to spare. Frequency got back into the Caddy and just drove around. Frequency drove around and just so happened to see a car that looked familiar. Frequency saw that it was Doc's brother-in-law Jay and he had hit a pothole. Jay was sitting on the hood of the car waiting on a tow truck driver.

Frequency pulled over and got out of the car to offer help. As soon as Jay saw Frequency it was as if a burden had been lifted off his shoulders. Frequency gave Jay a hug and then made a phone call to Webb. He told Webb where he was and what he needed. Webb told Frequency that he was in luck because he had a truck not too far from him. And true to his word the tow truck driver showed up. Frequency told the driver to take the car to the shop and he would handle the formalities. The driver said cool and left. Frequency told Jay to get in the car and he would take him to where ever he was going. Jay said he was going home because he was through for the day. Frequency told Jay to call his wife and let her know that he was alright and riding with him. Jay made the call and told his wife that everything was alright and that he was getting a ride from Frequency. Jay got into the Caddy and said nice ride. Frequency said thanks, but it is not mine. Jay laughed and said I should have known. Frequency cleared things up when he told Jay that this was Don1 car and he was on the way to pick him up. Jay apologized, and Frequency told him it was ok, he was not going back to the old days.

    Jay then asked Frequency what he was doing, and Frequency told Jay that he was just thinking about that. Jay was confused so he asked Frequency what was on his mind. Frequency told Jay he was thinking about starting a business. He just did not know what he wanted to do. Jay asked him was he interested in the computer business. Frequency told Jay that he owned part of EoW's computer shop and that would be a conflict of interest. Jay corrected Frequency by telling him about the computer consulting business. Jay had already established a company, but business was stagnant. Jay said he needed a partner that he could split the profits with evenly. Frequency asked Jay what his role would be. Jay said whatever you want if it will bring in the legitimate business. Frequency laughed as he pulled up to Don1's house. Frequency parked the car and then got into the back seat. Out came Don1 looking like new money. He got into the car and introduced himself to Jay. Jay told Don1 his name and then asked Frequency where they were headed. Frequency told Don1 that he needed some transportation because he could not be riding around in his Caddy all the time. Don1 told Frequency that he could not buy that rust bucket TA they saw last night.

    Frequency told Don1 to take him to Mr. Rob's Adult Toy Store. Don1 looked back at Frequency like he was crazy. Frequency assured Don1 that he would buy something civilized and nothing over the top. Don1 said sure you will and then he headed for the Toy Store.

Mr. Rob saw the Caddy pull up and knew he was going to have a good day. As soon as he saw Frequency his thought changed to how much it was going to cost him. Freq shook hands with Rob and told him that he was looking for a car that was as comfortable as that Cadillac but would not stand out in the hood. Mr. Rob thought for a moment and then he told Frequency to follow him. Mr. Rob showed Frequency a Hummer H2 1st, but Frequency said that it would draw too much attention. Mr. Rob then showed him a Jaguar, but Frequency said it is a FORD. Mr. Rob then showed him an Impala, but a police man drove by in one just like it. Mr. Rob gave up and said that he did not have anything else that fit the bill.

All his other cars were high profile cars like Mercedes, Range Rovers and Maserati. Frequency said no to all of them until they saw Jay get into a black Chevy Suburban. Mr. Rob told Jay to wait because that was his personal ride. Frequency looked at Don1 and Don1 told Mr. Rob that he said anything on the lot was for sale. Jay got inside the truck and immediately fell in love. It had all the makings of a mobile computer showcase. Frequency asked Mr. Rob for the key and he hesitantly gave Frequency the key. Frequency got inside and took it out for a spin. He liked it a lot and decided to buy it. Frequency went back to the Adult Toy Store and asked Mr. Rob to sell him the Suburban Mr. Rob tried to play horse trader with Frequency, but it did not work. Frequency ended up giving him blue book value for the truck. Frequency sealed the deal when he told Mr. Rob he would wire transfer the amount to him within the hour. Mr. Rob told Frequency he had to go get the extra key and the owner manual was in the glove box.

Frequency signed the contract for the vehicle and then wired the money to Mr. Rob. Frequency then told Don1 to tip Mr. Rob for making their shopping experience so pleasant. Don1 gave Mr. Rob $2000 and said that they would be doing future business with him. Mr. Rob smiled and said anytime just call; no matter if it's twelve o'clock midnight just call. Frequency told Don1 thanks and he would catch up with him later.

Mr. Mario sat back in the back of the "Galleria" high on the pills that the GayMack had done up for Ms. Bobbi. Mario did not know how to control himself. He had been in control of the "Galleria" ever since Ms. Bobbi left the country with the GayMack until the heat died down. The GayMack was recovering in Malaga Spain. He had been shot up in the fight with the FEDs but was able to escape with the help of his main thang, Ms. Bobbi. Mario wanted to go and get himself a new piece, but he had messed things up and raped one of the largest gangs' member in the city. To make matters worse it was the leaders' little brother. Mario selected his victim by how low they sagged their pants. Whenever Mario saw a young guy sagging, he got aroused and then he attacked them and fulfilled his urges.

Mario was scared, and he wished that Ms. Bobbi was here to clean up his mess. Mario saw one of the workers that he had been watching. Beana had just gotten back to work from her breast augmentation surgery. Beana was so glad to have her new breast and was not paying attention. When Mario snatched her into one of the storage rooms and started to molest her. Mario snatched her bottom away and her secret was exposed to Mario. Beana was born a man going through a sex change and he was trying to keep it a secret. Mario didn't care he just wanted to be satisfied. He spun

Beana around and entered Beana hard. Beana let out a scream but this storage room was sound proof. Beana fell to the floor and Mario fell on top of her and pounded her anus until she submitted to his will. Beana spread her legs wide and let Mario have his way. Beana knew that if he got his way, he would be able to make it out of this room. Beana found herself getting into the assault to the point he felt his own manhood come to life. Beana started pushing back to meet Mario's thrust. Mario really started to pump Beana and Beana made sure he went in deep with each thrust. Mario released his grasp on Beana's arms and he moved in a way that shocked Mario. Beana turned over on his back and pulled his legs up toward his head. He guided Mario's hard member back inside of him and Mario went deep. Beana pulled Mario's head to him and stuck his tongue in his ear.

Mario was turned on even more and he made Beana release a squeal that signified he had her. Beana was surprised when Mario pulled from her and started consuming her. Beana could hold back and orgasmed. Mario consumed her and then started to kiss her deeply. Beana swallowed and returned the favor. Beana gagged as Mario presented him with more than he could consume but that did not stop him. Mario could not get the satisfaction that he needed so he pulled from his mouth and went back to his behind. Beana grunted as Mario went for broke and pounded him twice as hard before. Beana totally lost it and told Mario he was his for the taking. Mario released inside of Beana and collapsed on top of him. Beana was amazed at how much seamen Mario released within him. Mario pulled from Beana and Beana laid there motionless. Mario stood over Beana and told him to clean him. Beana looked up and then moved stiffly to do as he was told. Beana licked Mario clean and then he put his member away. Beana looked at Mario and he slapped him in the face. Mario then told Beana that he needed to go and get cleaned up for work.

Beana got up and walked out of the room. Beana went straight to the dressing room of the club and cleaned up. Beana ached all over from the assault from Mario. Beana had made plans to not have sex until after he had finished all his surgeries. Now that his plans had been changed, he figured he better get his act together, if he wanted to keep his new suitor happy. Mario went back to the main office to find that he had used all the pills that were left behind by Ms. Bobbi. Mario was stumped and began to get agitated. He made some phone calls to a few local ecstasy dealers, but none of them had the pills like the ones he had been consuming. Mario was out of control and he could not believe how addicted to these pills he had become. Mario then heard a knock at the door and it was the manager of the "Galleria". He said that a package had come, and it was for him. Mario asked the manager what it was, and he just shrugged his shoulders. Mario screamed at him to bring the damn thing and the manager hurried off to go get the package. The manager returns with the package and left it with Mario. The manager closed the door behind him and went to get the girls ready for tonight's show. The manager went into the dressing room and made sure that everybody was on task.

He saw that Beana was moving stiffly and slow. The manager asked her was she ok and she said that she had a little incident and she was sore from it. The manager accepted the excuse and then went on to some of the other girls. Mario opened the box and inside was a cellular phone. He turned on the phone and the screen had a picture of a belly button. Mario did not understand at first then suddenly it hit him like a ton of bricks. The phone suddenly rang, and Mario answered it immediately. On the other end a familiar voice asked Mario did he miss her. Mario was overwhelmed to here Ms. Bobbi's voice. Mario immediately started telling Ms. Bobbi everything that had taken place since she and GayMack had left. Ms. Bobbi told Mario not to worry because she and GayMack were on their way back. Ms. Bobbi told Mario the full details of their arrival and what he needed to do to prepare for their arrival. Mario started writing down all the details and then he told Ms. Bobbi that he would ensure that everything was perfect when they arrived. Ms. Bobbi then hung up the phone and Mario sat back in the big office chair and reveled in the thought of knowing for sure that Ms. Bobbi and the GayMack would be back running things.

Jay arrived at Webb's shop and picked up his car. Webb did a full brake job on the car and put a new set of tires on it. He also gave the car a full tune up and had it washed. Jay was shocked at the sight of his car. He asked Webb had he heard from Frequency. Webb told Jay that Frequency was enroute to get some new shoes for the Suburban. Jay asked Webb could he see them, and Webb showed him the set of rims that were against the wall. The rims were 22 inches wide and were colored to match the Suburban. Webb then showed Jay the bull bar and the supercharger to increase the performance of the truck. Frequency pulled into the garage with BDB close behind him. BDB jumped out of the big CXT and told Webb that he needed to have his truck overhauled Webb said for real? And BDB stated that he was just bullshitting. Webb started laughing and then he got the crew started on Frequency's new ride. The crew tore into the Suburban and had the rims on it within the 1st 30 minutes. Next, they started putting on the bull bar. That took about another thirty minutes. The whole time they were adding the new rims and bull bar. Webb was getting the Suburban ready for the supercharger.

Webb had prefabricated something special for Frequency's Suburban. The supercharger that Webb had chosen had been modified to come on and off with a push of a button. Webb got the ideal from watching a Mad Max movie. Webb changed the tow button on the Suburban to an actuator for the Supercharger. BDB asked Webb what Frequency would do if he needed to tow something. Webb stated that Frequency would call him. BDB started laughing and then stopped and said that shit ain't funny. Jay interrupted and said that he could setup a diagnostic computer that had GPS and Navigational software built into it. Webb liked what he heard and asked Jay how soon he could get this system. Jay said that he just needed to make some phone calls and then he would have it shipped. Frequency asked Jay to use FedEx overnight and Jay said he would have to check. Frequency then told Jay to take him to the store because he needed to get a few items. Webb knew exactly what Frequency was about to do, but he kept his mouth shut. Frequency and Jay headed to the store and Frequency had Jay call about the system for his truck. The company was called Omnionix and they had a state-of-the-art system with no financial backing. Frequency had Jay to order his system and told them he needed by morning. The clerk on the phone said that he would have to check with the owner to find out if they could get a system there by tomorrow.

Mack Master was at the forensic lab trying to find any evidence from the samples given by the victims. He could not find anything, and he tried everything that the Fed's databases had to offer. Mack Master then had a crazy thought and made a phone call to Frequency. Frequency and Jay had just finished ordering the new system for the Suburban. Frequency answered the phone and Mack Master immediately began telling him his findings. Mack Master then asked Frequency could he use CMAX to help him find something. Frequency told Mack Master that he could connect in via voice command over his PDA Smartphone. Frequency made a conference call into CMAX and setup Mack Master to be able to have access. Frequency ended the call and let Mack Master and CMAX get to work. Meantime, Jay and Frequency picked up a few items at the store then headed back to the shop.

Mack Master started talking in commands to CMAX via his cellular. CMAX acknowledged each question as he processes as he answered each one with information that was over Mack Master's head. CMAX had Mack Master type in an ip address that eventually gave CMAX access to the computer that Mack Master was using. Mack Master stepped away from the keyboard and CMAX took over. CMAX put several theories used by some of the best assault case file detectives result. CMAX told Mack Master that he could leave the lab and he would contact him with the results. Mack Master logged off the computer and left the lab. Mack Master then headed for Don1 cellular shop to get an upgrade on his phone. Mack Master took the scenic route and headed to a neighborhood known to be a local area for gay couples. Mack Master thought to himself that if the culprit came out of this neighborhood. It would mean problems for this neighborhood. Mack Master went by the Starbucks and got himself a cappuccino latte. Mack Master overheard one of the patrons say that they had been accosted by a couple of gang bangers looking for a rapist. Mack Master pulled out his badge and introduced himself to the patron.

Mack Master asked the patron where was he when he was accosted? The patron said that he was just outside of the coffee shop in the back-parking lot. Mack Master thanked the patron for the information and then he went into the back-parking lot. Just as the patron said there were two guys in the back-parking area walking up on anyone, who came into the parking area. Mack Master walked by them and they questioned him also. Mack Master acted as if he was a regular patron and then not shows his badge. The two members were firm but polite in their questioning techniques. They informed Mack Master that they were looking for a pervert, who was going around sodomizing boys. Mack Master acted as if he was unaware and asked the men how he could help. The two men said that they would always have somebody from their organization close by until this pervert had been dealt with. Mack Master said ok and then he walked back to where his car was parked. Mack Master received a call on his cell instructing him to go to Don1 cell shop immediately. Mack Master ran to the car and headed straight to Don1 shop. Once he got there, he was met by Don1 and EoW with the information that CMAX had found.

    EoW and Don1 were stunned to find no traces of any body fluids.

But they did find that all the attacked victims had the same gel residue within the attack area. Don1 told Mack Master the news and Mack Master told them about, how the gangs in the neighborhoods were starting to conduct their own investigation. Don1 stated that if they did not find something out soon, the gangs would find a culprit and deal with them in their own way. EoW said and that is one day I will not be in town and the shop will be on lockdown. The guys laughed as CMAX sounded over the computer speaker system that he found another match. The substance was a piece of polyurethane that came from a lining of a polymer. EoW had CMAX run cross checks with the different types of materials that used this type of polymer. The results were endless. CMAX found the polymer in shoes, shirts, skirts, swimwear, and costume clothing. EoW got on the phone and called his partner named LabRat.

    EoW told LabRat that he needed to find out how that a certain kind of polymer that is used in everything, be used just in a sexual nature. LabRat was going to say a condom, but that was the obvious and that was not it. LabRat asked EoW could he and Frequency come by and look at some of his new experiments. EoW knew that meant that LabRat wanted a reason to get out of the lab and hang with Frequency. EoW new that LabRat idolized Frequency and just wanted a chance to hang out with him. EoW promised LabRat that Frequency would come by and see what he had to offer. LabRat told EoW that he needed about 3 hours to get the results and then he would be waiting for them. EoW hung up the phone and Don1 lit into him immediately. Don1 told EoW that Frequency was supposed to be hooking up with you know who, today? EoW totally forgot that Jay was going to strand Frequency and his sister in law somewhere alone. EoW immediately called Jay and asked him, how his plans were going. Jay told him that his plans had folded due to the other person could not take off until Friday. EoW was happy. He told Jay to bring Frequency to the cell shop, because they had some business that they all could benefit from.

LongHaul pulled up to the cell shop and EoW, Mack Master, and Don1 all got in. EoW took over the steering wheel and headed for the location of the LabRat's new facility. Frequency did not have a clue of what was going on. Everybody kept telling him to just worry about the new business and they would handle the rest. Frequency had already started designing business cards and making plans for a new website. He and Jay would put together a design for a billboard to announce the services they had to offer. EoW followed LabRat's direction and found himself at the old fire station on the outskirts of town. All the men got out of LongHaul and went up to the door. A set of automated doors on the side opened and all of them went inside. The building fit LabRat's persona to the hilt. There were experiments going on in almost every corner of the building. Mack Master was impressed and asked EoW how long have they known this guy. LabRat stepped out of one of the glass incased research rooms and said for a while now. LabRat then walked up to Frequency and gave him a hug and thanked him for all that he had helped him with. EoW laughed and said now don't go get teary eyed on me. Frequency smiled and asked LabRat was everything to his liking.

    LabRat said he felt like a kid in a candy store and could not wish for more. Frequency called out CMAX and the computer replied "ACKNOWLEDGE". LabRat looked as if he just orgasmed in his pants. Frequency turned to LabRat and asked him for the information he had for them. EoW looked at Frequency and said he hated when he did that. And Frequency reminded him not to talk around CMAX, if he did not want him to know what he was working on. EoW started to laugh and then LabRat cut in and gave them the on the polymer that was found on all the victims. Next, LabRat came out in a suit that was made from this polymer. LabRat told Frequency that this suit is used by bio-technical facility as a deterrent to keep the lab workers from accidentally sticking themselves with a syringe. LabRat said that all the suits come from a company that caters to aids facilities and drug clinics, to protect themselves from an affected client.

LabRat went on to say that lately this company has had a rash of burglaries, which included several cases of these polymer suits.

Mack Master immediately got on the phone and notified the other agents of what they had discovered and immediately sent an agent to the company to do some investigating. Frequency told Jay that they would need to do some research on web-based surveillance and security for commercial and private properties. Jay pulled out a piece of paper and wrote down the ideal that Frequency just had. Don1 then stated that if Jay was going to continue to work with the crew, he would have to upgrade to a Smartphone. Don1 told Jay that as soon as they got back, he would hook him up. Jay just smiled and then he asked LabRat, if it was possible to setup tracking device within the polymer to see if they could let it lead them to the pervert. Frequency told CMAX to try and make that happen, if possible. CMAX came back with an answer within a minute. CMAX explained that by using the information from LabRat's finding, if they injected just one nanobyte into the polymer during the molding faze. The Nanobyte could be used to track the suits virtually anywhere. Frequency told LabRat that he wanted these suits made to fit Mack Master and Jay. LabRat said ok and then he had Jay and Mack master follow him into the research room. There he had them strip and get into two cat scan type machines.

LabRat activated the machine and it began to slowly scan both men from head to toe. EoW asked Frequency, why did he choose them two and Frequency reminded EoW that they had nanobytes already in their systems and he did not want to indulge that information just yet. EoW acknowledged Frequency and said he would not want to be the one to have to explain to LabRat, how they got nanobytes to work within the human body. Frequency and Don1 both laughed and then they waited for LabRat to be finished with Jay and Mack Master full body scans. Jay and Mack Master scans were finished and then the two men got dressed and stepped out of the research room. LabRat told the crew that he had sent the scans results into CMAX. Frequency told LabRat that they would be in touch and that he would be contacted by CMAX, if anything went astray. LabRat said ok and then he walked the crew out of the lab. Frequency shook hands with LabRat and told him that as soon as he got a chance the two of them would go hang out anywhere, he wanted. LabRat was elated and told Frequency that he would be looking forward to it. Everybody got back into LongHaul and headed back to Don1's main cellular shop.

Jay and Frequency sat down inside of LongHaul and put together all the necessary paperwork to get their new business up and running. Jay really enjoyed the ideal of putting a computer consulting firm together, with a known ex – computer hacker. Frequency just liked the idea of being in a legitimate business, period. The two men hashed out all the details and then Mack Master to notarize the documents. After all the formalities were taken care of, Frequency broke open a bottle of Muscat wine, from the V. Sattui Winery Company. All the men got a glass and toasted to the success of Frequency and Jay's new business.

## 3.) Back 2 Business

Frequency walked into LongHaul and started running the data that Mack Master had gathered. The individual, who committed these crimes, was very careful not to leave any traceable evidence. Frequency picked up a file with the name Rev. Michael Hypocrite on it. Mack Master entered LongHaul and said long time no see. Frequency turned to greet him and say, "Yeah had to scratch a pesky itch for business". Now what is the deal on this guy? Mack Master stated that Michael Hippocrite is a pastor of a self-made church. He was raking in millions off the backs of religious people. The part that was suspect was that Rev. Hippocrite had several members missing. All the members were underage, and their parents' tithes had fell short from their normal amounts. Frequency asked Mack Master had the church membership grown since the people came up missing. Mack Master said he had not thought of that Frequency told Mack Master to upload what he got into the system and he and CMAX would get on it.

    Michael Hippocrite sat back in his office and just took in the view. He had the best of security installed so that he could see every crevice of the building. He was so glad that he took over the "Galleria" building when Richard Inman had to leave town suddenly. Never in Richard's mind would he imagine his club turned into a religious cultural center. Michael Hippocrite couldn't imagine calling his concept a church because of the stigma that went along with that name. Michael however used all the financial advantages of a religious organization to increase his profits. Michael was looking into all the security monitors when he saw a vehicle pull up to his private drive and then pull into his private garage and park next to his vehicle. The driver got out of the car and let a woman out of one side. He then went to the trunk of the car and pulled out a wheelchair. The driver then went to the opposite side and helped the other passenger out of the car and into the wheelchair. The three individuals headed straight to the private elevator and activated it to take them to the Michael's office. Michael turned on his personal security system in his office and grabbed the special remote to activate it. Michael made a special note to get his security people back out to recode all the private drive and elevator security system.

Ms. Bobbi stepped out of the elevator 1$^{st}$ and then in rolled Richard Inman pushed by Mr. Mario. Mario was still coming off the drugs that he had been taking, which left him not as sharp as he used to be. Michael walked over and greeted his surprise guest. He had them go over to a sofa and chair area and asked them to make themselves comfortable. Michael then started to explain to his underground investors how he established himself as an "elder" in a church that grew to have a fairly large congregation but a "difference of opinion" with the head pastor of that church led him to be released from service... he then decided to get his religious cert. from online seminary schools and professed himself as Pastor of his own ministry - Restored Original Covenant ministries better known as "ROCm". He used the ROCm as a secret tax shelter and means of inducing the poor community to flock to his false offerings of worship, praise, and spiritual enlightenment.

He used the influence of Judge Warren Hill to get financial backings to help get the ministry up and running. In turn the Judge filtered his money and henchmen into the church to keep the membership in line and continuously growing. Before the Judge went to prison, he would make the young outlaw gangsters go to the ministry to do community service. His people on the inside would recruit the youth and get them involved in the illegal activities when the judge needed muscle. Now that the Judge is locked up, Rev. Hippocrite is taking over where the Judge left off. Hippocrite is taking care of his handy work by using some of the Church members, who just so happens to be a part of a gang that Hippocrite reformed into his Security staff. The same individuals on the street was called the "ROI" short for Remove, Overcome, incapacitate because that is what they did to anyone that got in their way. Rev. Hippocrite called them his Religious Organization Investigators to keep it on the Up and Up.

Now Hippocrite got involved with the GayMack and Ms. Bobbi before he had discussed the matter with the Judge. The Judge still had killers on the street to handle his business, but two of them the Judge tried to double cross. The plan backfired and now they were squeezing Hippocrite for money after they found out he was tied in with the Judge. Rev. Hippocrite explained the process of how he extorted money from the members of the ministries and how loyal his staff was. Richard Inman applauded the good Reverend and then told him what he expected from him. Reverend Hippocrite walked to his desk and then activated his office security system. The lights dimmed to show his guest the lasers that surrounded the area that they were sitting in. Ms. Bobbi reached to grab her weapon, but a laser pointed straight to the middle of her forehead. Michael spoke aloud to ensure that his captive audience could hear him. He told them that any movement was a death move. He told them that they were surrounded by a laser activated trigger system that had gun with silencer built on them. Any movement by anything contained inside would cause the whole group to be annihilated. Mr. Mario was scared straight. His senses were fully aware and all he could do was stand there.

Mr. Michael then told his guest that they needed to watch as he put on a show for them. The time had come for Rev. Michael Hippocrite to do one of his three daily broadcasts from his desk. The area around his desk lit up and Reverend Michael Hippocrite started his religion-based broadcast. Michael brought forth sincerity and humbleness as he spoke the divine words. He put forth passion in his heart as he reminded everyone that he could not help healing the sick alone, but he needed their help and divine giving. He then ended the broadcast with a reminder that Christ paid for our sins and we also must pay for the healing forthcomings. He then asked that everyone send in their offers to a P.O. Box or go to the Ministries' website and make an offering using, check or charge. All cards are accepted and if you cannot get through keep trying because Christ did not give up on you, so you should give the same in his name. The lights went off and the good Reverend immediately went to the financial tabulator and watched the donations come in. After he had seen enough the Reverend turned his attention back to his guest and told them what he was going to do for them. He told Richard Inman or the GayMack that he was going to pay him rent for the building plus 25% of the judges take on the organization.

The Reverend then shutdown half of the containment area and told Richard and his crew that they had one minute before the field would activate. GayMack said he understood and then he and his crew moved quickly out of the office and back down to the Gaymack's automobile. As soon as the elevator doors opened the ROI was there to escort them back to their car and off the premises. As soon as the car left the grounds, GayMack stated to Ms. Bobbi that he left this guy with a "MINT" and all he wants to pay back is rent and twenty-five percent. Ms. Bobbi saw that the GayMack was really pissed and asked him did he want her to take care of this for him. Richard looked over at Ms. Bobbi and told her to get a hold of the Madame. GayMack stated that he wanted to catch this fool and make him pay for what he had just done to him. Ms. Bobbi pulled out a phone and was soon talking to the Madame.

Madame was just finishing her workout in the gym, when her phone rang. She listened intently to everything that Ms. Bobbi had to say. Madame told Ms. Bobbi that she would be back in the states as soon as possible. Madame had given birth to twin boys and was raising them in Italy. She had all the legal documents put in place to send the boys to their father, if anything should happen to her. Madame missed Big Trouble, but new the consequences of seeing him again. Plus, she had been on the run from the Fed's thanks to the Judge. Now Madame had a chance to redeem herself and get some of the money owed to her from the Judge's crew. Madame let word with the nanny that she was going out of town to take care of some personal business. She told the nanny that she would be gone no longer that 2 weeks. Madame got all the necessary items she needed to travel to the states and left to catch a plane to the states. Madame knew that this trip would bring back old feelings, and she made it up in her mind not to give in to them. Madame enjoyed her new-found freedom and lifestyle in Italy. She had learned the language and had a large stash of money to take care of her for a long time. Madame thought about the information that she got from Ms. Bobbi. Madame knew just what to do to turn a square out. She would use whatever it took to get the good Reverend a calling. Madame was soon on a flight to America. She had all the credentials falsified thanks to the GayMack and she had no trouble making it in. Madame walked into the airport.

Frequency Hunter and LabRat had all the communications in place. Frequency wanted to get all the data locked and uploaded for the next evolution of nanogentech.

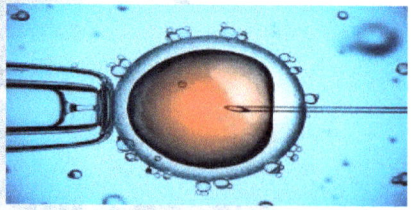

## 4.) Fireman

BOOM

The heat was intense, no sign of any movement, the entire hangar was engulfed in black smoke and flames. The whole building had been destroyed in one mega blast that set everything in it and around it in flames. Alarms were blaring from car alarms and surrounding buildings. Nothing was moving in the area until the distance, siren came on the scene. The firefighters immediately jumped into action trying to contain the massive explosion. The 1st responders immediately put on oxygen tanks to go search for anyone that might have survived the blast.

The area was sunken and filled with heavy debris, no heat signatures or lifelike structures seemed present. The 1st responders backed out and allowed the rest of the team to put the fire out. As the Lt. of the fire team arrived on scene, things got a little tense. He was about to call out to get some information about the building and suddenly all the communication equipment shutdown. Lt. couldn't contact the captain to notify him of the assessment. By the time, they were about to start troubleshooting equipment, a set of SUVs' came speeding toward them. Overhead a large helicopter flew over and blew smoke away from the area. As the firefighters continued to fight the blaze, the helicopter touchdown and some black dressed individuals stepped from the copter. Next, heavily armed men and women jumped from the SUVs and immediately locked down the area. The Lt had seen this before, it was something to do with homeland security. The Captain finally, was able to get through and was notified of the black out in communication. He was informed that agents were on the scene. The Captain informed the Lt. he was enroute and would make the scene shortly. The agents spoke to no one they just setup shop and started to assess the situation. As soon as Capt. Arrived another agent stepped from the helicopter. She was very beautiful but stern looking, she addressed the Captain and told him her name was Agent Tryce and this facility belonged to one Mr. Frequency Hunter. Captain acknowledged her and then asked, if he was terrorist or bombmaker? She said in some circles, that would be one of the things he would be called, others he would be called a savior of young souls. Captain looked confused, so the agent told him how Mr. Hunter brought down one of the biggest pedophile rings and shutdown an entire child trafficking network. He had made enemies throughout the world and so we thinking somebody just wanted a little payback. The Captain stated if you call this little, I would hate to see anything else.

Agent Tryce then asked if any casualties were found, the Captain stated now we couldn't get close enough to check fully. Over the radio, sniper announced a fast-moving car was headed to the area. Agent Tryce told them to stand down, it was probably agents from the local field office. The car skidded to a halt, agents Mack Masters and Layrock got out with a look of desperation on their faces. They both immediately stated that Frequency was in the building when the explosion happened.

Don1 was playing one hellava game of golf. He was just about to tee off, when he heard some unfamiliar sounds coming toward him fast. He remembered what happened the last time he waited to see what was going on, so he immediately ran for his golf cart, by the time he was about to get to it a four-wheeler landed on the golf course and bee lined straight for him. Don1 was cut off by the armed rider but one good swing with his golf club caught the rider directly in the jaw knocking him off the 4-wheeler. That gave Don1 enough time to get to his golf cart. He jumped into the seat and immediately took off on a cloud of air. Don1 loved his hydrofoil golf cart that his wife gave him for Father's Day. He would kiss her and the baby later, now he had to just escape, whatever the hell was going on. The attacker got up and saw that his target was getting away, so he called for backup. They arrived quickly with other motorcycles and atvs equipped with machine gun firepower. Don1 heard them shooting at him so he went full throttle on the hovercart. The attackers were gaining on him and were getting close to trapping him until Don1 drove into the pond. The attackers stopped their direct attack and went around the edge to try and capture him. Don1 moved across the water like it was frozen.

The attackers just fired shots, but he was out of range. They pushed the atvs hard to get to the other side where they saw Don1 get back on land. The attackers knew they could catch him now because some of their guys had went and stolen some boats. Don1 had been playing scenarios like this on video games so, he just had to play hard now that his life was on the line for real. He immediately, pointed the hydrocart to the river on the other side of the 18th hole of the golf course. He figured these guys probably would have did some homework on the course but wasn't expecting the toys Don1 was using.

Don1 figured somebody finally wanted to try to get even as his wife had been saying all along. Don1 made it to the big lake with the attackers trailing close behind him. He hit went across the water with no problem what so ever. He lifted a protective cover on the dashboard and the hovercraft transformed into a battle-ready machine. Don1 remembered when Webb customized the craft while he and his new wife was away on vacation. The regenerative solar battery and water turret setup was amazing. The hover cart could run forever with the power of the sun and water. Don1 went full throttle because he knew he had room to run. He immediately sent out warning codes to his wife and his crew to let them know what was going on. Don1 reached under the seat and put on the helmet, that was the best part of the hovercraft. As soon as he put it on the heads-up display, started uploading information to Agents. Don1 didn't understand why CMAX didn't respond, but now, he couldn't dwell into it. As Don1 rounded the bend shots peppered directly in front of him. The attackers had caught up to him and were firing high powered weapons at him. Don1 activated defensive mode of the hovercraft. Basically, a small high-tech version of two miniguns shooting 40cal rounds.

Don1 was able to simultaneous aim and fire the weapons using the heads-up display. He shot the guns in burst to save ammo until help arrived. The attackers were savvy, but they didn't do their homework. Don1 was able to anticipate their movements and counteract them as if he was playing a game of speed chess. Don1 remembered the protocols of the military program that was used to make the heads-up display, so none of the attackers' vehicles were shot above the keel line. He mainly left them dead in the water for the authorities. Don1 was holding his own until one of the attackers got a shot into his electrical system. The hovercraft lost one of the propulsion units. Don1 moved into survival mode, he shut down the firing system, and used the power he had to maintain speed. At this moment he could really use CMAX. He kept his vehicle moving just enough to keep out of range of the weapons of the attackers. Next thing he heard a helicopter bearing down on him. He was about to shut down the hovercart, when he heard the authorities announce shutdown your vehicles and drop your weapons. Don1 looked up into the sky and thought "Kiss the Baby".

The fireman contacted his sources and stated the drops had been a success, but a vehicle had escaped the assault. Mynt shutoff his communication with the fireman and went to give Roz the news. Roz let Mynt know until Hunter was found no one against them was safe. Mynt laughed at her and stated that Jahoodi was after her head and she need to tell him how to shutdown Hunter's entire circle. Roz told Mynt that she gave him everything that Frequency allowed her access too. Now that the full assault had been put in place, the two of them had better get somewhere off the grid until they found that truck of Hunter. Mynt didn't understand, so Roz explained to him how Frequency linked his entire circle from the inside of that truck. Mynt immediately called Jahoodi and gave him all the info that Roz had spilled to him. Jahoodi told Mynt that he would be there personally to pick them two to join him on his new "FoF" (Fortress of Finance). Mynt yelled at Roz to pack up everything because they now have a new locale to make home.

The plot to destroy the F.R.E.A.K.S. has been activated.

## 5.) Dump Truck

Peeaye Steel knew he had a lot of work in front of him. He started recycling the steel he had sold Frequency when he bought the Hanger. He sat in the construction trailer outside of the location running heat and seismic scans all over the area of the old hanger trying to find any clue, if Frequency had survived the explosion. Peeaye knew his old friend had a knack of always being many steps ahead in any situation that may befallen him. Peeaye inspected every large chunk of metal that moved making sure Hunter didn't leave sign that he needed help. Peeaye collaborated with Random Trucking to move all the steel products from the site, because he knew Random had close ties to Hunter also. If Frequency could get a message to any of them, he knew they would make sure no one other than their crew would be notified. Random had finally moved all the big beams and framing structures. He got into the crane and started pulling out the major piping joints and collapsed plumbing lines.

Both men were busy trying to ensure they were scanning every piece trying not to miss a thing, when

Cornbread called and told them about the details of the property they were working on. Cornbread was the crew's lawyer and they called him cornbread because he was from Mississippi and was raised the old fashion way. Cornbread enjoyed doing business with the crew purely, with the knowledge that he was going to make some serious money from whatever they brought to the table. The purchase of the hanger was right in line for Cornbread to finish paying off the deed to his law firm. It was not much but it sure beat the stress of working for a larger law firm with a lot of egos. Cornbread asked the men did they have any luck finding anything. Both men stated no but they were not stopping until every inch had been search thoroughly. Cornbread told the men to keep him informed and if they needed any money, all the finances were in place to take care of them, and that he would see them later and then ended the call.

Both men went back work trying to clear the site of all the debris from the explosion. Random's Crane operator suddenly alerted him that a pipe was sealed shut at each end and it would need some extra cutting to free it from the area that it was buried in. Peeaye overheard the message on the radio and immediately called out that he would be there to help. He jumped into his white pickup truck and headed to the area where the crane described. He surveyed the damage and just shook his head. Peeaye activated the lights and the area that the two men stood in lit up. Peeaye stood in awe at the massive pipe and the damage that had caused it to rupture and seal at both ends. Because of the type of steel of the pipe nothing was pinging from any of the scanners to show if anything was trapped inside. 3D images of the pipe inside could not be gathered by the industrial stuff that Peeaye had on site. Peeaye then told the men that he would get the obstacle out of their way and let them get started getting this enormous pipe set free. Peeaye told Random's crane operator that he did not come here to work, but to supervise. The two men laughed as Peeaye went inside of the pipe damaged area and started working out a way to free the big pipe.

Mack Master had just finished talking to Layrock about Frequency's disappearing act. Mack Master talked to Layrock about a hit out on a Federal Prosecutor. Mack Master and LayRock discussed how the prosecutor was working a case against a bigtime minister, who was using his influence to allege himself with some very unscrupulous characters that were known to break the law at their leisure. This minister had somehow taken over and made himself religious gangster kingpin. The minister had his people take out any obstacle that got in his way and now they were going to overstep their boundaries. Mack Master asked Layrock was the prosecutor in question called the Nightshade on the streets because she could poison any defense attorney's case. Mack Master was curious to find out more about this Nightshade and told Layrock that he would be back to work in two days guaranteed.

More cutting of wires and cables took place and then finally the big pipe finally was free from the collapsed building. The crane got it lifted out of the hole and placed on the ground for a platform for better inspection. All the guys were trying to figure out how did such a huge pipe get crushed closed on both ends. The decided to send it to a large federal holding facility until it could be cut open and investigated. The groundskeeper foreman called in the extra wide dump truck, so that they could fit everything inside to save cost. As soon as it was clear to move into position, the driver of the truck parked the dump bed perfectly so that the crane operator could lift and put the pipe in it, with all the other pieces to maximize his payout. As the pipe was dropped into the bed the driver recalibrated the scale to show that the dump truck could hold more payload. A few of the workers threw some damage large capacitors and batteries in into the dump truck and then another drop of pipe and wires. The driver was satisfied with his payload then he strapped down everything and started locking down everything so that he could get this load on the road. Everything was secure, and the money load was about to be paid.

Doorshaker was loving this connect he had with this private company. They paid him as soon as the truck was locked down with no question asked. This dump truck ideal was not his, but it had paid him and his best friend 3x the initial investment, they paid for it. Doorshaker drove the tractor trailer rig out from the premises and onto the main road. He let got out of the city and in no time allowed the big rig to do, what it was made to do, Haul ass. Doorshaker was enjoying the scenery and the music inside the truck took his mind off his troubles, until one of the alerts sounded from the trailer. Doorshaker looked at the alert and it showed as a tail light failure. Doorshaker knew that was not possible, but he did allow those guys to throw a bunch of other stuff in the dump bed. Doorshaker started to slow the truck down and prepare to move over to the emergency lane, when suddenly street jackers came into view behind him. Doorshaker immediately stepped on the gas and got the big truck back up to speed. Doorshaker had been hearing about these steel jackers and he was not about to be a victim. He had a contract to collect on and this one was a big payload. Doorshaker laughed at these idiots as they passed him in a SUV with some guys with pistols and shotguns. Doorshaker already had the rig configured to take small arms fire, so he was getting himself prepared for some trucker R&R, and that's ramming and running over shit. The guys in the SUV was preparing to fire their weapons, when they noticed that the truck had a video capturing system. Before they could pull away, the driver swerved to miss an animal in the road and was clipped by the speeding dump truck. The animal flew out of the way, just in time to not be hit by the SUV. Doorshaker laughed out loud and then he noticed more were coming. Doorshaker knew this was not right, why so many jackers were coming after him. Doorshaker needed some help and had to think fast, he was driving the truck hard when he noticed the brake light indicator kept going in and out, plus making a sound to match. Doorshaker tried to use his radio but there was no signal. Now he knew something was very wrong. Doorshaker knew he had to get to a rest stop or a weigh station. Doorshaker knew he had rigged all the information so that he could pass through without any problem, but it would see that he was excessively speeding. Doorshaker went for the weight station and knew they would help him. He went full throttle headed straight for the station. Meanwhile, the jackers had received their payments to steal this truck and get rid of the driver. They knew that they had to work with all the vehicles that they had, because there would be no backup and room for error. Streetjacker 1 headed to the weight station and killed the officers inside. As Doorshaker made it to the station, he saw the carnage that had been dealt to the officers. Doorshaker knew he had to call in the road dogs.

Doorshaker had a satellite phone network installed on a drone that followed the rig everywhere it went. He sent out the call and across his speaker a voice he was not expecting to hear.

EoW answered the call and knew exactly, what to do. He took over the drone program and started assessed the situation. Layrock and Mack Master were already enroute and they had plenty of firepower with them. The Authorities were able to catch up to all the action and they found out these guys were heavily armed. The firefight was intense and the dumptruck kept on a rolling on its destination. Eow asked Doorshaker did he need anything else. Doorshaker asked EoW was it anything that he could do about that damn brake buzzer. EoW laughed and started to shut down the buzzer, then he got quiet. EoW screamed to Doorshaker take this load to the container drop zone. Then the satellite signal went silent.

Doorshaker knew something was up with this load. He saw the navigation system change and the rig went into full stealth mode as he pulled around the corner and out of view of all that was chasing him. Something big must be going on with this load and suddenly, the brake indicator stopped and his favorite song "Purple Lamborghini" started to play. This payload just got serious for the Dump Truck.

# 6.) Thunderhead Hawkins

Cutter and Chopper Hawkins were known as the C and C Murder Machine. The two men had spent 4 years in each branch of the military service. They became a part of the elite forces of each branch before leaving and going into another branch just to say they did it. The two brothers ended a 20-year run in the National Guard and then they just disappeared.

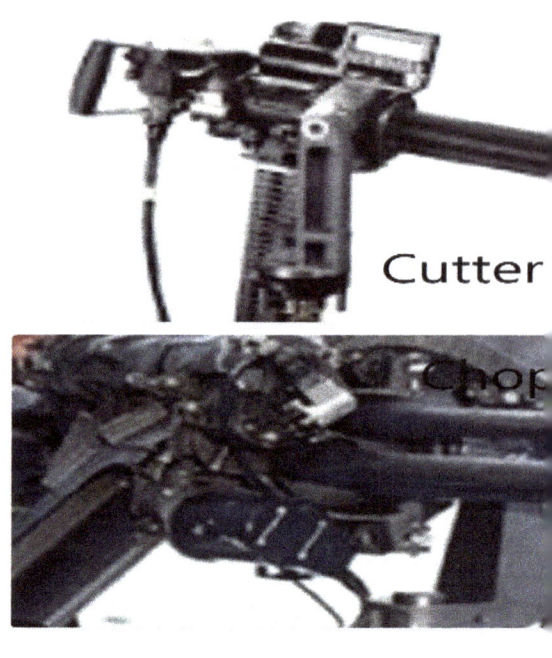

Cutter

Chop

They had a way of staying in touch with their father and that was about it. The two men never called home just sent postcards from various places they chilled at.

Thunderhead Hawkins had his own hideaway down in the Louisiana swamplands. It was a miniature fortress that kept the riffraff out and the animals close. Thunderhead would hear from his godson from time to time. He would send Thunderhead the latest tech in video equipment and would have the most up dated satellite TV system installed. Thunderhead was shocked to see his security monitor light up on the old path gate system. He knew instantly something was wrong because only 3 people knew of the gate system and two of them were out of the country. Thunderhead ran to his room and grabbed his fancy phone the boy sent him, but he hardly ever used. As soon as he pushed the button, the picture on the front display let him know, who it was coming in through the old path. The bad thing was the truck in the picture did not have any bullet holes in them. Thunderhead grabbed his 357 revolvers from the nightstand and went to his Cord to go investigate. Mid stride Thunderhead stopped and made a U-turn back to his bedroom and then down in to the floor of his closet. He put in a code and the door opened.

Thunderhead reached inside and grabbed one of the mini machine guns that were stashed inside. He cocked the H&K and then ran for his Cord. By the time he made it to the door the big black truck was pulling up the drive. Thunderhead ran outside and got behind his old Cord to give himself the advantage over whoever was inside.

The driver's door opened, and a man fell out of the truck. Thunderhead ran over to help the man up and it was his worst fear. He held the battered torso of his godson.

LabRat looked into the eyes of Thunderhead and felt the weight of the world lifted off his shoulders. He tried to speak but the wound to his face kept him from doing so. Thunderhead helped his Godson up from the ground and got him into the house. Thunderhead immediately called an old girlfriend of his that was a surgical nurse at the local hospital. He gave her explicit instructions and told her to come alone.

     Ms. Kathy showed up to the house later that night and immediately started working on LabRat. She removed all the debris from his body and was amazed at how LabRat was hurting. Thunderhead in the meantime was

rummaging through the truck looking for any information to tell him "What happen to his Godson"?
Thunderhead couldn't find nothing until he went into the back compartment of the truck and saw a small glow coming from the back glass. Thunderhead touched the glass and a computer-generated voice sounded. It said that the system was down to its last 2% of power and needed to be powered up as soon as possible. Thunderhead was clueless, he did not know where to start? CMAX came online and sent a text message across the screen. It was a question that only Thunderhead knew the answer too.

The question was 'When your back is against the wall, who do you call'? Thunderhead said "Family!!" CMAX initiated the charging system by releasing a set of cables from the bottom of LongHaul. Thunderhead then received a small printout of instructions on how to charge the system. Thunderhead followed the instructions exactly. Thunderhead had to rig up a pair of jumper cables from the main house power box and connected them to the cables from the back of the truck. Thunderhead then went back inside of LongHaul and tapped the screen as instructed. Inside the house the lights started to dim. And Thunderhead watched in amazement as the Big Truck came to life. All kinds of robotics and lights started moving about as if he was in a Star Wars movie or something. CMAX came online and stated that all maintenance systems had been restored and were initiated. Little robots were all over the place repairing the damages sustained to the vehicle. Thunderhead asked the question of what happened to his Godson? CMAX's instructed Thunderhead to sit down on the couch and he activated the monitor. CMAX then replayed all the events leading up to LabRat running to his godfather for help. Thunderhead knew his godson was good, but now he needed help. The people, who did this to his godson was not finished and Thunderhead knew they would come looking to finish the job. Thunderhead knew he could not handle this on his own, and this time he would have to bring in the C & C. Thunderhead was about to get up and leave when CMAX activated LabRat's monitoring system. Thunderhead didn't understand how, but he knew that this truck had something going on that was way over his head. CMAX told Thunderhead to bring LabRat back into the truck and put him in the bed. Thunderhead was just about to ask "What Bed? when a bed came from out of the floor in the back of the truck. Thunderhead smiled and went to go get his godson. Ms. Kathy was just finishing bandaging up LabRat, when his godfather came in and told her to help him move LabRat to the truck. Ms. Kathy didn't hesitate and helped Thunderhead get LabRat to the truck. As soon as Thunderhead laid LabRat on the bed CMAX initiated the RozWell program. A little robot came out and started injecting LabRat with a getwell serum and applying electrical charges to them so that they could start the rejuvenating process immediately. The debris wounds to both side of his face had been sutured closed by Ms. Kathy. The robot immediately injected the reconstruction bots into his face to get the damages repaired.

LabRat's godfather was amazed. He knew of the inventions and computer programs that his godson had been working on, but nothing of this magnitude. CMAX instruction Thunderhead to go get some rest because he would be having some guest by morning. Thunderhead said ok, as he heard a sound of electric shock and then a groan from his godson.

The Following morning Thunderhead got up and did his usual routine. He moved around slowly until after he got his shower. He then went into the kitchen to make himself some breakfast, but the scent of bacon and eggs hit him in the nose. Like an old bloodhound on the trail he went into the kitchen to see Ms. Kathy in one of his shirts cooking up a breakfast fit for a king. Thunderhead went inside and sat down to eat. He and Ms. Kathy talked about old times, when they were interrupted by CMAX. CMAX told Thunderhead that he would be able to speak to his godson shortly. Ms. Kathy looked at Thunderhead and asked him How was that possible? And Thunderhead replied they don't call it technology for nothing.

Thunderhead finished eating and got dressed. He and Ms. Kathy met at the door and went inside the truck to see LabRat sitting up and watching tv. Thunderhead went to his Godson and gave him a huge hug and a kiss on the cheek. LabRat spoke softly but legible. He told Thunderhead he was so glad he was there for him and thanked him for getting CMAX back online. Thunderhead then asked LabRat, what was this they were in and what is going on. LabRat explained everything to his Godfather not leaving out any details. He even showed him the footage of the two federal officers that were working with him, get shot up. Thunderhead remembered his manners and introduced Ms. Kathy to LabRat. LabRat thanked her for helping him and tried to give her some money for her help. Ms. Kathy told him she could not except the money and smacked Thunderhead on the butt and said that his Godfather will take care of that later. LabRat smiled as Thunderhead walked Ms. Kathy out to her truck and she soon left. Thunderhead came back into the truck and then asked LabRat what he needed from him. LabRat told Thunderhead that he would need some serious firepower because the man that was after him had connection with some Mercenary's' out of Columbia. Thunderhead told LabRat that he did not have to worry about that. Thunderhead told LabRat all he needed to do is get word to his sons and his mercenary nightmares would be put to an end. LabRat smiled and then grimaced as the pain went through his face. Before he could move the little robot rolled over to him and administered a shot that eased his pain. Thunderhead was amazed at the technology in this truck. Thunderhead said to LabRat that he had one nice truck and LabRat told him the trucks name is LongHaul. Thunderhead laughed and remembered some of the things he taught LabRat as a boy growing up. He taught LabRat that when you give a machine a name it makes it personal and for some reason it seems as if the machine holds up longer. Thunderhead smile was cut short as his thoughts ended abruptly. Suddenly, the sound of something powerful was overhead.

Thunderhead went to go get his guns, but LabRat stopped him. LabRat told Thunderhead that was some of the help that he was waiting on. Then Thunderhead heard the sound land in the water like a boat of, but this was no ordinary boat. This boat has some serious horses and it was coming up from the rear of the house.

LabRat told Thunderhead to go out and meet the fellas and to bring them to LongHaul. Thunderhead left and came back in no time with EoW and Don1. Both looking as dapper than a GQ magazine. Both guys were apologetic about not being there for LabRat. LabRat told them that it was not their fault.

The men, who did this to them had some serious jamming equipment. LabRat said the one thing they did not anticipate was the horsepower that was inside of LongHaul. As soon as LabRat said that Don1 had CMAX run a full scan over the truck and sure enough there were tracking devices in a couple of places. Don1 and EoW went outside and pulled them all off and showed them to LabRat. LabRat did not have to say a word, Thunderhead spoke up and said I need to bring in the boys. Don1 and EoW turned to Thunderhead and asked How could they help? Thunderhead told them about his 2 sons and How he only needed to get a message to them and they would come a running. EoW asked Thunderhead if he had an ideal where they were. Thunderhead said Cutter is probably somewhere in Cuba, and Chopper is more than likely in Jamaica off on one of those islands. EoW told LabRat to get better while they cleared up the details on how they were going to find Thunderhead's sons.

Thunderhead said because they did not know them, they could be in some danger because his boys was not playing with a full deck. Don1 laughed and told Thunderhead that his godson didn't fall to far from the tree. They laughed about it and Don1 started working with CMAX to get their communications back online.

EoW started hooking up the custom hitch to his boat so that he and Don1 could strike out as soon as possible. Don1 figured out what the Merc did to jam their communications and then designed a way to knock theirs offline and send an ultra-high frequency through their own device that would enable it. Don1 gave LabRat and Thunderhead new Smartphone just for this mission. He had CMAX sync everything inside of LongHaul, so they had a means of recording everything. Thunderhead was confused until LabRat broke all the technical mumbo jumbo into plain English. LabRat then told Thunderhead to move LongHaul to a place more secure than the front of the house. Thunderhead told LabRat that he tried but the thing would not respond to nothing he did.

    LabRat smiled and called CMAX. CMAX came online and asked

How could he be of assistance? LabRat told CMAX to give Thunderhead full rights to LongHaul. CMAX stated confirmed. LabRat then told Thunderhead drive it like it was yours. Thunderhead went to the driver's seat and sat down. LongHaul came to life showing Thunderhead all that was special about this vehicle. Thunderhead smiled like a kid in a candy store, as he revved the big truck's engine. Thunderhead put it in gear and started to drive LongHaul cautiously $1^{st}$, then he got more aggressive. LongHaul responded like a champ and gave Thunderhead more than he had anticipated.
Thunderhead kept on driving and grinning as he pulled off onto one of

many roads that ran deep down into the swamp. He found a spot where the boys use to bring girls to get their jollies off and knew that they had all kinds of boobytraps set.

LabRat was resting in the house, when he got the digital image from Don1 and EoW showing him the new helicopter and boat merge into one, and it was called the Airfish. It was an awesome sight to see and LabRat new there would be some money to be made when those patents hit the military table.

Thunderhead returned from hiding LongHaul. He was so proud of his Godson but scared also. LabRat looked over to Thunderhead and told him that everything was going to be alright now. LabRat then told Thunderhead to put on his headphones and watch the screen. Thunderhead asked why? LabRat told Thunderhead that he needed to familiarize himself about the new technology he was about to be involved with. Thunderhead nodded then stepped back to watch the guys takeoff.

**Airfish**

## 7.) Bad Man

Yahoodi Ibraham: **Made $6.7 billion as the kingpin of his criminal cartel organization. Anyway, there's not much to say about this maniac that isn't illegal. Sex trafficking, sports point shaving, illegal gambling jury tampering, bribery of federal officials, money laundering, drugs and**

## blowing stuff up. And that is just the stuff on record.

Yahoodi Ibraham was enjoying his new man-made island, that they named the "FoF, short for Fortress of Finance. His business took a loss of a big freight package to Frequency Hunter, due to the crash of the sex trafficking network. Yahoodi was sending mercenaries to get his freight back and to eradicate any and all peoples that had any dealings with Hunter. The mercenaries were the Somali Snatch Squad or 3S. 3S had one specific rule, grab the subject and triple tap everything else. They had been operating for years under the Somali regime, but after a few of the water raiders got killed, they decided snatch and kill were more profitable. The leader always did the negotiating and dealt directly with the squad. No others could talk with them.

    Jahoodi enjoyed the 3S work so he had no problem how they handle the business. Jahoodi only concern was to get back his revenue stream and see his competitors crushed. As for Frequency Hunter he wanted to test his skills for strategy in the high stakes gambling arena, where you walked out with the money or you die.

The trafficking network that took so many years to setup got compromised by a computer geek, nothing more. He was able to gather all his compadres and their personal finance information then used it to destroy the network. Majority of the weak links had been assassinated during the court trials. A lot of them were killed in witness protection or on the way to testify at their hearings. The 3S was taking out judges, jury, DA's and police officials. They had no qualms about making an example of any official that thought his title made him safe. Jahoodi really enjoyed hearing about bestial rapes of men and women just for the sake of a fear tactic. It made it easier for him to move in and out of the cities without little notice. Jahoodi had financing setup for many street gangs to help him keep up what was going on in the areas where he made his investments. Jahoodi decided that since some of his old constituents were stepping back into the game, this time they would get to meet the man that controlled the money and would inevitably, be the destroyer of Frequency Hunter.

Yahoodi was putting together the strategy to get rid of Hunter and his crew of freaks, when the leader of 3S was escorted to him to discuss how they were going to be able to get his men past the authorities. Yahoodi explained to the leader how his men would go into the port at night while his movie company would be filming a movie. The port authorities will be laxed because they will see that your crew are actors working on the set. Jahoodi made sure that a few of the men could speak English and knew the territory good enough to explain their situation, without killing anyone. The Leader smiled and asked what about the groupie bitches, the two last and Jahoodi said bareback their women and make good strong Somali babies. The Leader smiled and stated that we have no problem with. Jahoodi then had all the necessary vehicles ordered for his brief stay in the states. He had to make sure everything went as planned because as soon as the government realized he was on US soil, they would be sending out the best of the best to apprehend him and this was not going to happen without making a lot of Americans becoming seeds of the Earth.

    L1, was the ruthless leader 3S. He and his men were about to enjoy the fruits of their labor as they had all the finance needed to invade the US territories. L1 wanted to teach the US that they were not as strong a resilient as the Somali Pirate Regime. L1 and his men sat back for years learning different languages and dialects, so that they could infiltrate anywhere in the world, that their financer needed them. L1 Dropped a lot of his Muslim upbringing to become a very powerful tactician. He enjoyed seeing the women squirm as the men snatched, raped and killed all the people they assaulted. L1 knew getting into the US was a big deal, but no one from his homeland would imagine him enjoying the hospitality of a billionaire businessman.

A week has passed and now the 3S full team are on US soil. They have laid low and kept quiet of their intentions. Most of the men began finding good jobs as masonry and millwrights on construction projects. The others were taxi drivers and paint color specialists. The best of them hired themselves into a surveying consulting industry. These men were the deadliest of the 3S. They knew how to circumvent and bypass every communication system within the city they operated in. L1 knew all he had to do was wait on the word from Jahoodi and they would bring down the 3S assault on the unsuspecting infidels.

Jahoodi finally had all the players that were back up and running on the network. He knew he had to remind them that he was bigger than the Judge and that he would not tolerate any pig shit from any of them. Jahoodi decided to kill on sight any one of these people, just to show them that he is not taking any chances. Jahoodi then got word that there was a couple, who was moving some new kind of sex addictive drug. He didn't care who they were, he knew he would buy all they could make and then disperse it out to his group of guys to triple the income revenue. All of that would come in time, but first he had some big business to attend to, that would draw a lot of attention to everyone that's in the illegal trade business. Jahoodi grabbed the remote to his new practice drone. He flew it around the FOF and enjoyed the beauty of his amazing achievement. The island was beautiful and powerful. It had all the newest stealth radar technology and it had an infinity anchoring system which means they could anchor anywhere and weather any storm. Jahoodi really enjoyed knowing that he didn't have to refuel this thing ever thanks to the reactors purchased through some black-market dealings. Jahoodi just needed to figure out how to make this reactor a world conqueror instead of just an infinite power source for a man-made floating fortress. Jahoodi decided to save that thought for another day and concentrate on getting his shipment back from Hunter. Jahoodi laughed as his drone picked up a couple of kids having sex on the back of a watercraft. Jahoodi thought about sinking the vessel but then that would bring to much attention to the area. He guided the drone away from the area then landed it back on the FoF. Jahoodi had one of his servants come to him and told her to contact L1 and tell him, that he needed the best drones and pilots that he could find. It was time to send a message with a bang from the Bad Man!

FoF

United Kingdom high stakes gambling fortress

# 8.) Switch On

IraLyn Baton got out of her custom H3 with a chip on her shoulder. Her plans for an intimate night with her husband was destroyed by a football game and company. She had forgotten it was football night and her husband quickly reminded her that he was hanging with the fellas. Not to be a spoiler she let him have his night and put all of her pinned up aggression on her court case. IraLyn used this aggression on a regular basis in her court cases which earned her the court nickname of the "Baton", after a small but lethal bat with a bashing offense. IraLyn went into the briefing with her staff to be updated on her next victim. The case was high profile in nature due to the assailant had deep rooted ties with underworld organizations. The judge over the trial was one, who had his own problems and was in the underworld's pockets. He was in his study receiving a gift from his underworld ties. The young woman was giving the judge an exercise in fellatio that he was totally engrossed in. The Judge was snapped to reality when the bailiff knocked on the door to remind him of what time it was. The Judge immediately stood up and tried to regain his composure. The Judge went into his private bathroom to wash up before he took the bench. Unknown to the Judge the young woman left his quarters through an access panel in the floor and climbed down into a tunnel that lead directly under his courtroom bench. The tunnel had been put there years ago to use as an escape route for the Court Justice in case of an emergency. The assailants' underworld ties found out about the tunnel and decided to use it for another purpose. The young woman was told to do whatever it took to ensure that their associate got off clean. The young woman prepared herself as she saw the Judge come in and was seated at the bench. The Judge sat and watched as the DA and the defendant's attorney worked the case. The defendant was one Mr. Ray Edwards. Mr. Edwards was on trial for numerous charges of sexual assaults and pedophile activities. Mr. Edwards attorney had plenty of motivation to get his client off. The underworld had sent a couple of hitters to monitor the case. Mrs. Baton was going through her deposition, when suddenly someone walked into the courtroom. The Judge looked up and was temporarily focused on this individual. The attorney for the defendant, suddenly asked for a mistrial. The Judge was pre-occupied with the young woman under his bench finishing what she started in his chambers.

The Judge maintained his composure and told the attorney that he was denied. He then had Mrs. Baton finish her case. Just when Mrs. Baton finished, another individual entered the courtroom. The Judge held his composure as he exploded from his under-bench attention. The young woman passed a note to the Judge that made him shiver in his seat. The note said that he should except the dismissal of the case. The attorney for Mr. Edwards looked into the eyes of the mystery visitor. The man stood up and then left the courtroom. Mrs. Baton also saw the man and knew she had been setup to lose. The Judge sat there and saw his whole career go to the dogs. The

Judge gained his composure once more and looked Mrs. Baton dead in the face as she pushed to take the trial to jury. The attorney for Mr. Edwards then came forward and told the Judge that Mrs. Baton and her team did not have anything to hold his client on and pushed to have the case dismissed. The Judge told both individuals to have a seat and he would go to his chambers to decide. The Judge got up and went to his chambers. As soon as he closed the door, he heard a voice that made him damn near have a heart attack. Michael Hippocrite sat in his chair with the young woman, who just finished pleasuring him, standing beside him. Michael Hippocrite asked the Judge, did he understand the message he was given. The Judge wanted to ask Mr. Hippocrite, how did he get into his chambers, but he was in a bad position to ask this man anything. The Judge accepted his fate and agreed to follow the directions of Mr. Hippocrite. The Judge told Mr. Hippocrite that he had no fear from him and Mr. Hippocrite told him he was a man of God and he only feared God. The Judge called Mr. Hippocrite and Hippocrite quickly interrupted him and told him to address him as Reverend. The Judge nodded ok and then the good Reverend left with the young woman in tow. The young woman gave the Reverend the digital camera with all her footage pleasuring the Judge. The Rev. laughed to himself as he secured another court justice under his wing. The Rev wanted the DA known as the Baton under his wing, but he could not find nothing to get her on. The Rev. was upset at his crew for not getting him any information on Mrs. Baton.

Whiteface stood over the dead bodies in Baton's residence. He laughed as his time for vendetta was now. He commenced to hanging all the men in the house by spider wire braided cable. He created all his instruments of death and now he was about to leave pure carnage in his wake. He hung Mr. Baton by the back of his throat. He sent a miniature harpoon through his mouth and then up on the wall he hung him. He threaded each finger in succession to make sure that each man, would allow the world to see their pain. Whiteface had one thing and one thing only to prove. He was going to bring Baton and her dirty bitches to agony. Whiteface knew how she operated, and he had already tied up the loose ends, so he knew Baton would run and he would enjoy seeing that ass shake in terror instead of pleasure. Whiteface left a note for all the Attorneys and Judges. "Imagine crying WHITE tears!"

Jahoodi made some major purchases but so far, his most intelligent one was his beautiful Russian Nuclear Scientist. Tavia secured all the old plutonium from the old satellite that crashed in the ocean. She had everything that she needed to get it into a secure shell and on its way to power up Jahoodi pop plate defense. She knew once this was put in place Jahoodi would give her his complete trust and freedom. Tavia needed to have her freedom to ensure she could show Jahoodi, she was worth every penny he paid, to get her out of the sex slave trade. Tavia laughed as she thought about how Jahoodi slit all the slave traders throat and then had sex with her on top of their blood. She gave herself to him willing and promised him to make him a very powerful man and feared by all. Jahoodi made sure he reminded her the day she failed him, she would drown on those words.

Madame had everything she needed to get herself established as big money mover in the area. She had blown her chance at meeting the Reverend Hipp at the last city hall meeting, so she decided to have a night out and attend a concert in the city. Madame made sure she had all her protection with her just in case trouble showed up. Madame moved through the entrance and to her balcony seat. She made sure she had a best view because she really needed to relax and enjoy her evening out. As the show was about to begin one of the ushers of the concert hall, came and asked Madame could a last-minute guest join her in the balcony. Madame thought no harm and so she gave him the usher the ok. To Madame surprise in walked in two big men escorting none other than the Reverend. Madame thought to herself that her luck was about to change, and the good church beacon light was about to give up the goodies.

Reverend Hippocrite sat down for the concert and let the last-minute shenanigans leave his mind as fast as the money for the bribe did. He knew he was going to have to pay for the last-minute sitting arrangements, but he was not expecting company. The good Reverend made sure his security detail stayed between him and this woman he had joined for the concert. The last thing he needed was a scandal based on something he had nothing to gain from. As the concert started the nights festivities was well worth it for the Reverend. He was thoroughly entertained and got an eyeful of a very beautiful woman that was sharing the balcony with him.

Madame knew what she was doing, but she didn't have a chance of touching the Reverend because his security would not have it. They made sure his eyes could see, but his hands couldn't *touch*. Madame decide to allow them all something memorable to end the night. As the show ended Madame stood to leave and bent over to pick up her purse. The split in her skirt let the men see that up under that skirt was all woman and not an ounce of cellulite.

    Madame knew she had their attention and knew the good Reverend would find a way to get in touch with her. One of the men from the security detail stood and helped Madame make her way out from the balcony. Madame thanked him and headed for the valet area. Madame noticed that the security for the concert had been tighten, so she asked the young girl what the reason for all the extra police was. The valet stated that some big shot had come to the concert hall and lately a lot of protestors been showing up because of him. Madame smiled and gave the girl her ticket to go get her car. The young girl waved to one of the other valets and he went and got Madame fully loaded 300C. He pulled up in the car and gave Madame the keys. Madame wasted no time getting away from the concert hall, but immediately noticed that she was being followed. Madame laughed because obviously this person must didn't know she had been going to tactical driving schools, while on the run in Europe. Madame lost the tail quick and the driver totaled his car as a result. Madame decide to go to a different hotel room, just to throw off the person that was following her. Madame paid for her new room and made sure that it had a Jacuzzi in the room. She made sure all her weapons were loaded and on point. Now all she had to do was get the jacuzzi ready and relax until daylight.

Reverend knew he had to have that woman, but he didn't think it would be that difficult and it would cost him a price for a new street car. He enjoyed the thoughts of mounting that woman before one of his security details got to her. This would be befitting of a hefty night bet and probably actually having to show her off at a church gathering or two. The Reverend laughed out loud as he got in his car, fastened the seatbelt and turned the switch on.

## 9.) N DA Game

And now Roz waited patiently for results of the new formula. She followed all the guidelines that had been put forth inside of LongHaul. She just needed the extra chemical makeup from Mynt. The bell rang as Mynt stepped into the process area. He told Roz that she did one hellava job stealing Eow findings from the RV's computer network. He then added a few addictive chemicals and some dissolvable designs and voila, the new high was born, and he had the exclusive patent. Roz brought the test tube to Mynt and he took a toothpick and gave a lick of it to a male laboratory rat. The animal immediately tried to mate with the female in the next cage. Mynt immediately gave the female the same amount and opened her cage. The animals immediately, began mating. Mynt had a camera on the animals and timed the event for accuracy. He knew he had just found exactly what he needed to bring the world to his doorstep with an unlimited abundance of cash. Mynt called to Roz, who was starting to leave the area. Mynt had her secluded and did not, want her to find out about the attacks of Hunter and his crew. Mynt needed Roz to stay close and help him finish the antidote for the new drug he entitled "Lick'em". Mynt finished the chemical makeup on the computer, then he asked Roz to verify his breakdown. As soon as Roz nodded yes, Mynt touched her lips with a toothpick tip size dosage of the Lick'em serum. Roz licked her lips and was confused, she asked Mynt why he give that to her without further testing. Roz immediately went to wash it off but by the time she rinsed her mouth, the serum had taken effect. Roz turned to Mynt and became insatiable. Mynt touched the toothpick to his tongue and allowed it to run its course. Mynt grabbed Roz and fondled and kissed her into the surveillance room, were they went into an all-out sexual frenzy, with no stamina or wanting issues. The experiment lasted for hours, with neither individual stopping due to sexual performance issues. Mynt just kept telling himself over and over, that he was in the game and the underworld would be his playing field.

50 million pills, 2100 pounds approximate shipping weight, with a street value of $1 billion USD. Mynt now had his biggest challenge, how would he keep Yahoodi from killing Roz.

Lick'em hit the streets and just as Mynt predicted he was moving major volumes. He kept the makeup of the serum between him and Roz. He constantly kept moving from different locations to keep them safe from the street thieves. Mynt kept them living under the radar until he could close the deal with Jahoodi. Mynt knew Jahoodi wanted Frequency and his hold crew in the ground, but he had other plans from Roz. Roz had always been his property, and no one could take her from him. He laughed at how many times she went to rehab and would return to him all cleaned up and educated, just to find herself back on her knees, drinking from his faucet.

Mynt had Roz risk her laugh befriending Hunter just to steal from him at his most vulnerable moment. Mynt now had information about the one-person Jahoodi would give a fortune for. Mynt knew Roz was the weakest link in any chain. Mynt thoughts were disturbed by a knock to the door. He looked through the peep hole and it was the next set of buyers. The 3 of them always showed up to buy the same amount of product, Mynt called them quarter, nickel, and dime. These 3 women were teachers at the local college. They loved to catch the youngsters slipping and use their influence to get their jollies off. Mynt laughed at the ladies because they always were trying to get him to come to their sex parties. He would always tell them no and they would laugh at him telling him, he did not know what he was missing. Mynt said, he wasn't into the kids and the one called Nickel stated that tricks are for kids and she was about to trick all month thanks to his "Lickem" pills. Nickel pulled out $5000 in a bank rapped note. The other two had their own payments in envelopes. Quarter had $2500, and Dime had a $1000. Mynt handed the ladies their product and escorted them toward the door. Dime dropped a flyer on the floor and Mynt reached down to pick it up. He saw the name Frequency Hunter on the paper and immediately stopped in his tracks.

Mynt turned to Dime and asked her how she knew Frequency Hunter. Dime stated that she never met him, but she teaches a class with a student that idolizes him. Mynt immediately, asked Dime does he comes to any of your parties. Dime stated that the kid was a total nerd. Mynt laughed, and Quarter asked is that the kid they called LabRat. Dime said yes, one of the little girls I fool with says he has Hunter on speed dial. They say Hunter treats him like a little brother. Mynt was totally interested in this kid now. He told the ladies to hold tight, he had something special for them. He went and got Lickem strips. Mynt told the ladies that he just finished this batch and wanted them to tell him, how they worked. Nickel started to try one, but Mynt told her, they were instant reactive. Dime laughed and told Mynt they were not trying to have a man, because they were into women, extremely young women, understand. Mynt nodded and open the door, he told the ladies, he would look forward to seeing them again.

Mynt immediately closed the door and began coming up with a way to track this kid without being picked up via any communication device. Mynt knew he had to keep this a secret from Roz until after he had what he needed to keep her alive. Mynt knew exactly how to play

Roz into thinking she was helping them. Roz knocked on the door and Mynt told her to get everything packed, they had to move again. Roz was frustrated, she was tired of running and hiding. She knew she messed up crossing her friends, but this was getting pathetic. Roz went to her bag and found a note that Don1 told her hold and forgot about. She unfolded the paper and it gave instructions how to temporary mask your communication signature. She called to Mynt and asked him could he do something with this? Mynt had an ideal, but he knew pf someone that could make it happen. Mynt asked Roz did she get them some transportation. Roz smiled and said we are now official members of the Gen Riders car club. Mynt went outside and Roz worked a deal with one of the Genesis Riders for an older model Hyundai Genesis coupe. They loaded up the car and went over the chop shop to get some electrical components to fix their grid problems. Mynt made Roz lay the seat back and get some rest, while he drove to the spot.

Mynt finished everything he needed to get Frequency on leash. He laughed thinking how Hunter would feel when he found out, his own little brother dropped the dot on him. Roz was ready to go when Mynt got in the car and placed a big kiss on her. Roz was lost, he never did that, she didn't know whether to get in position or dropped and please him. Mynt told her no more of that, you have earned your title, no more junky pleasures. Roz looked at Mynt as if a halo had appeared around his head. She nodded to him in agreement and asked him, what was next? Mynt stated that they were going to sell as much product as possible to get a major buyer to buy all of it and then they could go change the past and move into the future, without a care in the world. It's just me and you Roz, stated Mynt and nothing will take us back to the way we were. Tears started running down Roz's eyes. She put her hands in Mynt's hands and swore she would die before she did anything to hurt Mynt. Roz laid her head on Mynt's chest and he drove for the next hideout they would call home as a couple.

Roz loved their new crib, it had everything they needed to finish making the communication blockers and all the pills. Mynt allowed her to go out and have some girl time. She kept everything simple but classy. She really enjoyed hearing about the ladies, who were using Lick'em pills to get their freak on. Roz laughed when she heard about a female prosecutor getting assaulted because she was talking about cracking down on users of the pills. Roz just nodded and laughed, never commenting on the situations. Roz would leave and immediately walk to a shop or two before going to the car. She would always activate the communication blockers as she would enter or leave to make sure no one was tracking her back to Mynt. Mynt really enjoyed having Roz all to himself. She really made sure there was nothing he desired that she wouldn't try to please him. Mynt surprise Roz with a scooter to zip around on. Roz walked in and screamed with pleasure. She was so happy and now she had even more freedom to enjoy the area. Roz hugged and kissed Mynt. The alert light started to flash, and Roz turned to see a car come into the RV park. Roz knew that car she had seen it before and she didn't know what to do. Out of the car came Ms. Bobbi. Mynt smiled and said now let the big money makers pay. Roz told Mynt that woman is a killer and to be careful, she will fuck you up and then kill you. Mynt looked confused, so Roz bluntly stated that is a transvestite, with a big appetite for necrophilia. Mynt nodded and stated he would make sure everything would be purely business, no samples. Mynt kissed Roz and told her he would return. Roz ran to the bedroom and got the revolver, she was not going to be a victim of any of those sick individuals afraid of Frequency Hunter. Ms. Bobbi waited to hold a bag of money like she was instructed. Mynt appeared from behind of one of the RV's and stated that Ms. Bobbi was not expected. Ms. Bobbi stated that Mr. Inman was in the car and didn't want to fool with his wheelchair. Mynt acknowledged that and stated that he would be ok with him dropping the top on this fine automobile you are in. Ms. Bobbi told the driver to drop the top and he did. There he was the man himself the GayMack. he looked as if he was not doing well, but that didn't stop him from coming out to meet with Mynt. GayMack asked Mynt to excuse his seat, but he was not feeling as strong as usual. Mynt nodded and asked GayMack how can he help him? GayMack told Mynt that he wanted $50000 in those Lickem strips and he three of those tracking devices he made. Mynt nodded, ok and ask how soon you need them. Ms. Bobbi threw the bag over to Mynt. Inside was five bank rapped stacks of $10000 each. Mynt looked and stated you are man of your word. Mynt gave the GayMack driver a gps locator to where his stash was and gave his driver the 3 tracking devices. Ms. Bobbi ask Mynt, how they knew this product was good. Mynt laughed and said he enjoyed living and he was sure they had already experienced the power. So, no need to try and mess up a good thing. Ms. Bobbi said I guess you scared of me and Mynt stated not scared but don't mix business with pleasure. The driver hit the button to close the roof and headed to pick up the product that they so desperately needed.

Mynt didn't move until after he got the signal that they showed up and picked up all their product. Mynt went and got some clamps and picked up the money. After carefully microwaving all it he burned the bag in the place where it landed on the ground. Mynt then went back underground to his waiting woman. He got into the spot and immediately undressed and got into the shower. Mynt was overjoyed at the amount of money he just made. He felt Roz join him in the shower and he turned to see her fully dressed in a gun in her hand. Mynt was confused until she smiled and said I was all dressed for war and you came in with all that money, guess you must undress me for love. Mynt and Roz didn't need anything to help them at that moment. Roz made sure Mynt earned the sleep he was about to get because she was going to drain him and leave nothing to be shared.

The GayMack waited on Ms. Bobbi to finish with the informant that was in the trunk of the car. They had been given information that Mynt would not come close to them because of what happen between them and Frequency Hunter. The GayMack knew he was one of many that would like to get ahold of Hunter. His thoughts were interrupted when he felt a kiss from Ms. Bobbi, she was winded and exhausted looking. Ms. Bobbi told GayMack that that stuff was the real deal. All she did was put a strip in her mouth and give dude some head. Ms. Bobbi had to get him and herself off three times before she finally stabbed him in the back. The driver pulled away leaving the dead body bent over exposed as Ms. Bobbi had left him.

A few weeks later, utility crew found the body decomposing and unrecognizable. Meanwhile, Lick'em went from a taboo subject to the top headliner. Anybody and everybody were using it for pleasure and business negotiation. The power plays during business changed because so many people had dirt on others that the local prosecutors and judges, were afraid of doing their jobs because they didn't know who was holding what against who.

Mynt and Roz was holding majority of the cards. They had video footage of all the power players kink services. Roz was lost in love and Mynt felt a new type of power because he now had the girl, the cash, and control. He was now felt like the master in the Game.

## 10.) White, Black, and Red All Over

Ms. Bobbi and Mario were high as ever and they had another victim entangled in their web. They had Ms. Bobbi lure another young man out for a quick one-night stand. He had no clue it would be his last night period. Jack had never told anyone that he enjoyed sex with transsexuals. He stayed undercover and went out occasionally to have himself a fling or two, then he would return to his day job as an assistant minister at the church. Ms. Bobbi had worked him over good with her charm and Jack had to have her. What he failed to realize was that she had another with her that wanted him. Jack was overwhelmed by the pills that he and Ms. Bobbi had taken, that he was totally off guard, when Mario mounted him and took him down. Jack tried to fight but Ms. Bobbi had a firm grip on his manhood and testicles that he could not pull away. The three of them fell to the ground and Mario took Jack for own pleasure. Jack tried to fight and squirm in the beginning, but soon gave in to Mario's desires. Ms. Bobbi continued to please Jack with her mouth to the point Jack could not control his own body. Jack lost all control when his body shook with a powerful orgasm. Jack felt his body go numb and then all he saw was Ms. Bobbi in front of him holding something bloody. Pain suddenly shot through his body as he looked down and saw that Ms. Bobbi had removed his testicles and manhood. Jack tried to scream but Mario had run his blade across his juggler and was enjoying the pleasure of death on his body. Mario released Jack's lifeless body and went to Ms. Bobbi for a deep kiss. The two of them fooled around over the body and then cleaned up the area. Ms. Bobbi and Mario had stolen a car earlier that day and now they were going somewhere to dump the body. This was a ritual to these two ever since

Ms. Bobbi had return to the states. Mario was just happy to have them back. He preferred to be a follower than a leader. And that role fit him perfect.

## *White OUT*

    Mack Master never thought this day would come but the headlines that ran across his screen just solidified all the rumors that was on the streets. "Mr. White goes to Jail". Mack Master sat back and thought about their long drawn out confrontations and wondered, what brought the man down. Everyone knew Frequency had other ideas for him that had nothing to do with prison, but it seemed as the authorities had the notorious negotiator captured. White knew they would make a spectacle of his capture and hoped it would bring Hunter out of hiding. When he got into the folds of the police, he heard the shocking news that the body of his old friend hasn't been found and presumed dead. White wasn't believing anything until he was confronted by the ones that were closest to Frequency. No simple aerial attack could catch Frequency with his guard down, no this was someone, who had serious connections and the only way to get in touch with them is to let the authorities think they had him.

    Jahoodi was aware of who Mr. White was. He had his 3S on the way to make sure Mr. White would be his guest on the FOF. The authories were questioning Mr. White, when the 3S hit the holding area hard. They dropped frag grenades all around the building then dropped inside to start attacking officers. The 3S double tapped anyone in their path and taking hostages was not an option. They moved into the room outside of the holding area where Mr. White was being held then exploded the wall next to the fortified door. They executed everyone, but Mr. White then had him follow them out of the building into a vehicle, where they made a hasty retreat to meet with Jahoodi.

By the time the reinforcement got there, pure carnage was left behind. No one was left alive and the star prison had been taken away. Agent Layrock received the news and couldn't believe what he was hearing. The agency had been warned about this type of threat, but they didn't take the necessary precautions. Now a lot of grieving families would be left with a small pension and a bill to bury their loved ones. Layrock called Mack Master and they both agreed to meet in an undisclosed location to discuss today event. The two men knew they would be the center on what to do to bring an end to this, but they were missing the key ingredients. Mack Master told Layrock that Mr. White stated the only reason they were able to capture him was because he needed to know, if it was true about Frequency Hunter. Layrock laughed and said, how can we acknowledge something we don't know ourselves. Mack Master asked what we are going to do about all the other people that are left out here vulnerable now that these individuals have shown that they are not afraid of attacking anyone on their own turf. Layrock stated he wasn't sure, but he knew he wasn't playing by the rules anymore. He figured if this hit squad double tapping all victims, then he would shoot $1^{st}$ and discuss the death footage at his hearing. Mack Master agreed and the two decided to move Webb and BDB into a secure bunker to do some major upgrades to LongHaul. Once they came back out into the public, the team would have to be fully prepared for something equivalent to the zombie apocalypse.

The body count in the holding facility was unimaginable. These individuals killed anything moving including the fish in the tank. Blood was spread everywhere and Jahoodi loved the carnage. He was looking at the video footage and decided to have himself a little entertainment. Jahoodi called for Mynt and Roz to come into the room with him. He asked them both why should he spare their lives instead of feeding them to the 3S. Mynt stated that they still have unfinished business and that Jahoodi still haven't found Frequency Hunter. Roz was about to speak but Mynt cut her off by saying that he and Roz could keep the civilians at bay, until all Jahoodi's competition was buying from him. After all they were the masters of the "Lick'em" elixir. Jahoodi was about to shoot Mynt in the face, when Roz dropped to her knees and stated, we are your beck and call Lord Jahoodi, why kill the mere scientist that can bring you even more power and wealth, without the hassles of land-based governments and military. Jahoodi smiled and then asked Roz what she had in mind. Roz immediately told Jahoodi that she had stolen the elixir ideal and another ideal from Frequency Hunter and his close friend EoW. She told Jahoodi that there is a woman out there that has Frequency's heart and more than likely knew where he was. Jahoodi immediately told Mynt that he would live another day and he better thank his girlfriend for saving his life. Jahoodi left the room to await his newly captured guest. The leader of the 3S escorted Mr. White from the helicopter into the mobile trailer of Jahoodi. Mr. White didn't like being out of his usual clothes, but in this moment, he had no choice. Jahoodi walked in and saw White still handcuffed and in prison garb. He immediately told his staff to get him out of the rags and into something more befitting. After White had gotten cleaned up and changed, he and Jahoodi sat down to do some hard negotiating to get the business between them sealed. Mr. White was looking around and got a glimpse of someone that he knew. It was Roz and now she was with Jahoodi. One of the 3S took his mask off and asked White did he see something wrong? White immediately told him he would address it with Jahoodi. The man stood and told White to follow him to Jahoodi. Roz and Mynt was cleaning up from today's events, when they were summoned to Jahoodi's dinner and meeting room. As they walked into the room Roz was shocked to see Mr. White sitting with Jahoodi. As soon as White saw Roz, he told Jahoodi that Roz was working with Hunter's crew and they all were heavily into some high-tech experimental shit. Jahoodi laughed and stated that he expected to get the truth out of the weak Americans. White took a sip of his wine and said I'm not weak just a businessman that is after his best interest, which is him. Roz started to tell Jahoodi everything about all the experiments that she was a part of. She even told him about how Eow used their information to enhance his sexual girth. Jahoodi laughed and stated no way, at that time Roz showed him how Mynt and her manipulated the data to invent the "Lickem" elixir. Roz told Jahoodi that when she had the information that she and Mynt needed, she left because Hunter's crew was getting suspicious. Jahoodi looked at all the information that was presented to him, and then he called for the leader of the 3S. He spoke to him in Somali and then he addressed everyone in the room that, they better be working on a plan, because if this stuff works like Roz has described it, they would be the biggest epidemic since cocaine.

Mr. White was silent as he sat in his compartment awaiting Jahoodi and his decision. A very young girl came and told White to follow her. He did as he was told and followed her down some stairs to the outside of a semi-truck. Jahoodi's base was inside of a tractor trailer. White was impressed but it was quickly dismissed when he saw two naked young children hanging naked by their arms. White held his tongue because he knew this wasn't about to end well for somebody. Jahoodi walked into the room with Roz and Mynt. They had the new batch of "Lickem" in a syringe and Roz was going to administer it to whoever Jahoodi chose. Jahoodi stated that he was no Jeckel and Hyde, so he had the leader of the 3S to pick one of his men. The leader had the guards go get one of his men that they had under the truck in a cage to punish him. The stripped him down and cleaned him up with a water hose. They took a barrel of hot coconut oil and immersed him in it. They brought him into the area with the naked kids hanging. The area was a big cage that after the barrel was placed inside the cage was sealed so that no one could escape. The barrel was popped open and the man stood up from the barrel. The guys made fun of him because he was smaller than the child hanging from the cage. Jahoodi told the guys to hush and he told Roz to go ahead the experiment. The man tried to escape the cage, but he couldn't get out. He was so busy trying to escape that Roz was able to slip up on him and stick him with the needle. The elixir hit his bloodstream and into the heart. Immediately, you seen his demeanor change. He fell to the floor and he became erect. He stood up and started trying to grab Roz through the bars. Roz stepped back and then he turned his attention onto the hanging children. He used the contents left from the barrel to oil the children down then he assaulted them violently. He was like an animal with nothing but primal instincts to mate. This went on for a couple of hours nonstop. The children scream in suffering agonizing pain as this bestial assault came to an end with the death scream of the two. White stood there emotionless as Roz and Mynt kissed in celebration of their creation. Jahoodi turned to Mr. White and asked him now you understand why I am going to need everyone focused on the mission. White nodded as the man they used as a Guinea pig came down off the drug. He saw what he had did and looked down at his body. He screamed out, look what you done to me and tried to attack Jahoodi. Jahoodi carried a Turkish kilij under his tunic. As soon as the mad man came close Jahoodi beheaded him. His body released a spray of blood all over the white and black floor. The 3S leader screamed out warrior chant and the men began in cadence. White went back to the truck and into his compart where he puked. He couldn't believe what he had just seen and now understood why Hunter was trying to stop them. Those two little children were destroyed and then celebrated over. White knew he had to walk away from the money and do what was right, only problem was that Frequency was the only person, who could help him now.

# 11.) She Didn't Get It

Curves sat down on her bed backtracking the events in her head. She had one hellava roller coaster ride and everything seemed to be slowing down. Curves had met a wonderful man and had decided to settle down. He did not have a clue of her past and Curves was planning on keeping it that way. She had changed her name twice to hide the harshness of her past life. Curves left home for college and ended up working the streets for a Pimp named Shugga Dick. When she met him, his name was Richard and he was a four time

Junior in college. He always told her that his family paid his college tuition just to keep him away. That was furthest from the truth. Richard had been cut off from his family the day he arrived at the college campus. He totally survived and put himself through college by hooking up college girls with businessmen, at the local business convention center. Curves remembered the first day she met Richard. He walked in on her giving one of the geeks a hand job for the answers to their latest test. Richard waited till the dude orgasmed and then he burst in on them. The little geek jumped so hard that he squirted directly into Curves face. Richard had a video camera setup and he had videoed the entire scene. Curves immediately ran away, but Richard had her exactly where he wanted her. Curves hurriedly ran to class and Richard awaited their next meeting. Curves had no idea what was in store for her and it was not long before she found out. Richard took her three days later and Curves would never be the same. Richard was her first and he made sure she could not think twice about anything. Richard turned Curves out in less than a month. He had her doing any and everything under the sun with any Tom, Bill, and George, who had money to spend. Curves stayed with Richard until she graduated from college with a degree in Physiology. Curves was overwhelmed with gifts from her family and friends at her graduation ceremony. Curves mother and father came to her graduation and basically kept her with them the entire time. This gave Curves the time she needed to get away from the torture of the past years. Curves avoided Richard for weeks and managed to get away from the college without seeing him. She immediately found work and buried herself in it. She met a guy named Cal and totally absorbed herself into him to the point of taking on his last name, even though they were not officially married. Curves maintain her quiet peaceful lifestyle, until one day Hell rolled in a wheelchair at the local mall. Curves saw him but made sure that he did not see her.


Richard rolled through the mall to buy clothing and a few trinkets of jewelry. He had not brought a bunch of items with him, because of the size of the plane. Also, Richard had not planned to be in the states for long, even though, it looked as if his plans were changing fast. The business they left behind was lost in drug gang wars. The local gangs had nothing to fear so they went after territories left abandoned by the old regimen. Richard had already setup meetings with his old Dr. and was trying to find out how soon he could get up out of that wheelchair. He enjoyed being catered to hand and foot by Ms. Bobbi and their staff, but it was getting a little old and frustrating. Richard knew it would not be long before the authorities would come a knocking and with that Frequency Hunter. Richard wanted so badly to see that muthafucka at the end of Ms. Bobbi will, that he was willing to risk going to jail for life. Richard dismissed his thoughts as his driver told him that they had a meeting with his private physician in the old city morgue. Richard laughed because he knew all their connections had to go rogue due the Frequency Hunter and his crew of goodie boys. Richard told the driver to pay for his items and then go to the meeting. Richard put on his shades and didn't bother to check out his surroundings. As the driver came with the items and they headed out the door Richard thought he notice someone from his past. He didn't dwell on it because the last thing he needed was for someone to call his full name out loud.

The old morgue was just as bad as he remembered it from the past. Dr. Stamps was still doing street surgery. If you had enough money or EBT credit, Dr. Stamps was the man. He was a wealthy plastic surgeon until a high-profile person, went against Specialist Dr. documentations, and died on his table live on the air. Society scorned him, and the family ruined him. Dr. Stamps went underground and started helping anyone that needed or wanted a surgeon. Dr. Stamps would use the left-over organs and tissues to do experiments with. He would then turn his findings over to the black-market Doctor network. Dr. Stamps was famous in the trannie girl circuit because he could get the best Botox and silicone injects on a food stamp budget, hence how he got the name Stamps. Dr. Stamps saw the car pull into the tunnel and then back into the corpse garage. When he saw that it was Richard, he was so happy.

Dr. Stamps hurried out to meet him and his driver and knew immediately that Richard wasn't looking to good. Stamps immediately told the driver to get Richard in his operating room so that he could see what was really going on with him. Richard thanked Stamps and told the driver to make sure that they were not interrupted for nothing. Stamps immediately hooked the hoist up to Richard and stood him up out of the chair. Richard let out a painful growl, but Stamps hit with the good stuff to make him relax. Stamps used all his x-ray equipment to figure out how to help Richard the best. Stamps pulled some synthesized DNA from a fighter that had come across his lab and then injected it into all of Richard's damaged areas. Stamps then ran ultraviolet light across each wound and the muscle healed. He knew it would hold and keep Richard on his feet but for how long was the question. Stamps then gave Richard several heavy dosage steroid shots to help him use aggression to get past the pain process. He knew Ms. Bobbi would enjoy it because of her sexual nature. Dr. Stamps watched as Richards body started to straighten out and he was able to stand. Dr. Stamps then went and grabbed something Richard had been missing since they been on the run, his Cane.

Curves made it home and was about to tell her man everything, when he walked in and kissed her deeply. Curves kissed him deeply and he ripped her clothes open and started kissing on her nipples. Curves felt as if something had started to affect her as her man pleased her. Her climax came hard as he tongues lashed her. Whatever he had taken was now affecting her also. Curves didn't want to talk she just wanted to please her man anyway he wanted. Two hours had past and Curves awoke to Cal's snores. She got from under him and went to clean herself up. She had never had sex with him since the entire relationship. She didn't know whether to celebrate or cry. She turned on the water in the shower and just allowed it to relax her and wash the soreness away. She dropped the soap and bent down to pick it up she turned around and in the shower with her was Richard. Curves tried to scream but he covered her mouth. He told her if she screamed that man, she loves would be the latest victim of the serial rapist. Curves nodded and then Richard told her what he expected from her. Curves acknowledged that she would do whatever he wanted if he left Cal out of it. Richard told Curves that she needed to take Stamps all the information on the experimental psychological drugs that were just introduced. Richard told her if she took care of Dr. Stamps, he would not come back and pay her a visit. Curves braced herself because she knew that Richard didn't never turn down a naked piece of ass. Richard stepped out of the shower and took a good look at Curves. He then had to admit to Curves that when he 1st saw her, he thought he might want to hit it one time for old time sake. But after seeing her without anything on, he would let that thought die. Curves said what do you mean? Richard flat out stated that you have let yourself go and it's a wonder that old boy could fuck you even if he is on those pills. Curves was about to try and get and attitude but Richard's hand around her neck reminded her that she was still flesh, to a flesh peddler.

Richard and the driver laughed at how he was able to track down Curves. Richard told the driver he wished he would have known she looked that bad. Richard said did you see the gut, butt and back fat. No one wonder her old man was in the bar complaining about fucking her. Her body looked like the blubber librarian. The driver laughed out loud and stated did you see all the corsets, plastic wrap, bodysuits. She forgot how to look good outside of her clothes. Richard laughed and stated then they wonder why, the men with good eye sight, see them for what they are worth, an insurance policy. The two men laughed as they were headed for the next meeting. Richard thought about the past and how many of his old girls had totally became unimaginable. Just as he was into his thoughts the driver turned and parked the car. He gave the phone to Richard and it was Curves, she was cursing angrily and threatening to tell the authorities that Richard was back to doing old business. Richard laughed at her and asked her, have she checked on her old man. Curves ran to check on him and then let out horrifying scream. The phone went dead, and Richard dropped the phone out of the car.

The driver reached into the armrest and pulled out a new phone. The driver said, he just didn't understand, what was so hard to understand about the message. Oh well, he guesses "She didn't get it"?

PFV

## 12.) Quarter, Nickel, and Dime

Quarter, Nickel, and Dime was trio that had made a life pact with one another.

These 3 women were lovers as well as friends. They kept everything within their circle and would not leave a stone unturned when it came down their individual feelings. The women always called each other by their nicknames. Quarter real name was Quarterlynn Thomas, Nickel real name was Nichelle Bakker, and Finley Dime. They all hated their real names and stuck to just their nicknames.

Quarter took the lead of the trio, she was the main bread winner. Nickel kept the trio on the move by taking care of their transportation, and Dime ran the household of the trio. The girls kept within their circle and never allowed, what they did to get outside of it.

    Nickel was about to leave for the neighborhood store when she noticed an unusual vehicle outside of their home. Nickel walked passed the vehicle and kept walking to the store. As soon as she got out of sight of the vehicle, she called Dime, who was still in the house. Dime looked out of the window and saw the vehicle. She told

Nickel that she could not see the tags on the car and she could not see inside, because the windows were darkly tinted. Nickel told Dime to lock all the doors and set the alarm system, just to be safe. Dime said ok and she would call her if anything went down. Dime did like she was told and started to clean up the house. Nickel then called Quarter to tell her what was going on. Nickel was surprised when the phone call went straight 2 voice mail. Nickel blew it off due to Quarter told them that she had a very important meeting and that she would be busy for the $1^{st}$ part of the morning.

Quarter morning started as usual then it went way out into left field. She started working on a document for her director when she accidently hit a combination of keys on her computer that took her into a secure server. Quarter looked at the screen but did not think anything of it. She quickly exited the screen and went back to the document she was working on. Not know that her computer was compromised Quarter kept working to finish her assignment. LabRat finally got the opening he needed to get the information to help find Frequency and bring down the good the network of criminals. LabRat launched a worm program and was about to get out when he noticed that the company had a video network linked to this secure server. LabRat knew he had struck gold with this information and setup a cloning program that sent an encrypted file of all the video feeds directly to the FREAKS temporary server under the classified area of this investigation. LabRat ensured that he had full access to all password and logins to this server so that he could use it later if need be. LabRat knew Frequency had built up a distrust for the good Reverend and made a promise to be there when they took him down. LabRat pulled up the video feeds and watched all the surveillance cameras feeds. He watched as the security guards started their rounds then split up. LabRat didn't think anything of it until one of the guards started going into the different offices looking for something. The guard are not supposed to have entry codes to the executive offices, but this one did. He went inside the office of one Mr. Nagel. The office was very nice it had all the accessories an office could. The guard sat in the office chair and then pulled something out of his pocket and then the video feed went blank. LabRat immediately went into action and got the camera to come back online. It did not take long for LabRat to get the camera back on line and what LabRat saw took him for surprise. The guard lay in the chair dead. LabRat tried to get the camera feed to play back when he notices something going on in another office. A young woman was working at a computer and she was about to get up when suddenly there was an explosion in the office that sent everyone into chaos.

Quarter had the big diversion she was waiting on. She went had all the information changed over and now, her crew could finish up moving more influence through the school. Nickel had made it back into the house when all the commotion took place. Her and Dime started getting ready to vacate the premises and live the life of Amazon queens.

Moufpiece was adjusting to her new place of refuge. She was kept under watchful eye of the LMG as they went about disrupting the sex traffickers drug ring. PPK would check on her from time to time just to make sure she was still aware that she was under his watch. Moufpiece would watch the news as the feds were raiding and showing up everywhere looking for Frequency Hunter and the attackers at the center. The news crew would always interview the same agent Mack Masters and he would always say that they were working to bring this madness to an end. Moufpiece didn't understand how deep she was really in and didn't care. These women had been using her as their pleasure piece since she was eleven years old. Moufpiece had always catered to them, one at a time or all together, high on drugs and alcohol. Moufpiece could remember the 1st time her brother and her got paid for the sex and told to keep it a secret.

    Her brother would always say, they enjoyed having her spread apart with her hands fully inserted inside of them as she used her tongue skills on the leader. Moufpiece could render the women into a pleasure paralysis with her hand inserted deep inside of their pleasure core. The one called Quarter rally allowed Moufpiece to fist her because she didn't enjoy being in a submissive state. Moufpiece would always have to go down on her quickly and furiously before Quarter would try to choke her out. Moufpiece knew when and where to get to her and that made Quarters knees buckle every time. Moufpiece would always get something put down her throat for that, because Quarter didn't like losing her power to her little pleasure pup.

Moufpiece thoughts were interrupted by LabRat knocking on the door. He saw that she had tears in her eyes but knew not to get involved. His mom had been moved to a secure location and Thunderhead was just as stubborn as ever. He knew he would be ok and something about the F.R.E.A.K.S. were being spread that they had been annihilated by a new drug cartel. LabRat came to talk to Moufpiece about any info that she had. Moufpiece smiled as LabRat was looking as lost as ever. One of the gang members was glaring at him, like he was about to steal his lunch money. LabRat hurriedly stepped away from the guy and Moufpiece asked him what he was doing there.

    LabRat told her that the GayMack was back in business and he was on the prowl for the individuals that hurt Ms. Bobbi. PPK stepped into the room and told everybody to leave except for them two. PPK then explained to them that he had already been hit up for the reason Ms. Bobbi was attacked. He lied to them as usual because Ms. Bobbi was still unconscious and unable to describe what the attackers looked like, except Moufpiece. LabRat had the look of fear and Moufpiece didn't flinch. PPK stated that they think it's a vendetta hit on Ms. Bobbi because of what she did to your brother and others like him. Moufpiece had the acknowledgement she needed to kill Ms. Bobbi for the murder of her brother. PPK smiled and said C4 comes in small packages but, still does a lot of damage. LabRat understood exactly what he meant by that and asked PPK if it was ok, if he designed some security measures to keep Moufpiece safe, just in case things doesn't go as planned. PPK agreed and LabRat set off to go find one of his best gaming partners to cook up some serious tactical power.

# 13.) Teacher Teacher

LabRat was losing it. He is trying to hold it together, but Frequency Hunter is missing, and he was there when it all went down. LabRat constantly kept blaming himself but the guys would not let him dwell on it. His godfather was helping him communicate daily with Don1 and Eow. They were looking for his Uncles and they were making it difficult. LabRat was in deep thought when something caught his attention. It was a familiar sound. He cleared his thoughts and started focusing on where the sound was coming from.

He looked, and he almost fell out of his seat. His teacher had one of those drones flying over her head.

LabRat watched as the drone dropped something on the ground and flew away.

Ms. Dime was getting an emergency stash of Lick'em in for the school meeting. She needed some leverage to convince the school board to allow them to buy the houses next door for a new practice field. She already had the construction company, concrete trucks, and project managers. Ms. Dime had all these companies due to arrangements, with the owner's wives. She had video of them all with underage girls and boys. She didn't touch none of the males, but she couldn't keep the girls out of her mouth. She had a favorite one that they call Moufpiece, to be so young she knew how to keep a lesbian weak. Dime looked around to make sure no one saw her getting her product and headed straight for her class. LabRat was hiding in the corner and now he was curious, why was a teacher getting so much product delivered to school. LabRat waited a few minutes then he moved out into the open. He walked around and notice how some of the students were acting. One was Moufpiece. He walked up close to her and she was crying.

Moufpiece was trying to keep from crying but couldn't. LabRat pulled her to the side and asked her what was wrong? Moufpiece told LabRat they found her brother dead and discarded in the woods. She said he had been molested and then stabbed in the back. LabRat remembered everything from the files in LongHaul. It was a woman that did some other victims like that, he couldn't remember, who it was but he would not say it to Moufpiece. He knew she was a teacher's pet and she would go back and tell them everything. LabRat gave Moufpiece his condolences and stated, if you needed someone to talk to, he would be around.

PPK had every street soldier he could count on, in the streets getting info. He needed Frequency help but even he couldn't be found. That crazy attack on the Hanger would come with all its own set of repercussions but PPK just wanted the muthafucka who raped his brother. His little sister had called him and asked for him to come get her from school. PPK didn't ask questions he just had one of his captains drive him to go see, what was up with his sister. PPK made sure to ride in one of the low-key cars, due to all the heat in the streets. They made it up to the school and he saw his sister waving to him. PPK knew something wasn't right so he grabbed his pistol and an extra clip. He knew his captain had heavy artillery, but he didn't want to hurt any of the students. PPK got out of the car and went into the building where his sister was. She grabbed him by the hand and whispered to him to be very quiet. She led PPK into the empty classroom and then opened another door to where PPK could observe two students in the room having a serious discussion.

LabRat and Moufpiece was having a conversation about how to find out, who killed her brother and tell on the teachers that were blackmailing the board of directors. Moufpiece came up with an ideal to capture video the 3 teachers and use it to get information.

LabRat asked her how she would keep them from telling the police. Moufpiece stated that after the Board meeting the ladies were taking a leave of absence. They have setup getaway to keep the victims from getting the blackmailed information back. Moufpiece asked

LabRat of the silent drone experiment he had made to help the Disaster Recovery Dept. LabRat told Moufpiece that he didn't have the help of his friend Frequency Hunter. PPK hit the window and startled the two inside. PPK went inside the room and told the both not to move a muscle. LabRat was scared and Moufpiece decided if she was going out, he would have to do more than that. PPK little sister stepped in front of him and grabbed Moufpiece and pushed her against the wall. Moufpiece had the wind knocked from her, so she lost what little fight, she had. PPK told his sister, he just wanted to talk to the dude. PPK asked LabRat, how he knew Hunter.

LabRat stated he was his mentor. PPK asked him did he know were Hunter was? LabRat stated honestly, I am not sure. But he was sure he could find out a lot more if he could just snatch up one of his most hated enemies. PPK stated that Mr. White is still on the run. LabRat was shocked, he asked "Who are you"? Oh, he is my brother PPK leader of LMG. PPK then stated so you the smart one they call LabRat. LabRat stuck his hand out to shake hands with PPK and to his surprise, PPK shook his hand. PPK then asked LabRat why he was going to so much trouble to set up traps with these bitches. LabRat told PPK that, if he could get ahold to these three and one others. He could find

Hunter and put a stop to all this sick perverted sex attacks. PPK smiled, but something was missing. Where was the gold? PPK stated that the Gold Grilled Skull would not appear until the rapist of his brother was in his possession. LabRat understood, what he was doing and knew it was no turning back from the decision, he was about to make. LabRat went into his backpack and gave PPK a usb drive with a shopping list of all the items he would need. As for Moufpiece she was expelled from school and was now needing a spot to chill until she could get back on her schooling. LabRat told her that it was no way she could come to his house. His mom was already tripping about

Frequency being missing. PPK said that Moufpiece could go with him and she could get the chambers setup for their newly trapped guest. LabRat laughed and told Moufpiece not to get her ass kicked for messing around in any other business. Moufpiece licked her tongue out but couldn't say a word because PPK sister grabbed her by the tongue and had some very harsh words for her if she even had a thought of messing with any of the LMG members. Moufpiece nodded ok and she was given back usage of her tongue.

    PPK was about to leave, when LabRat asked PPK did he have any issues with the RhoidWreckers? PPK looked puzzled, and then he asked do you think they did this to my brother? No, LabRat stated, the police had all them checked out, but the Queen of them will have the answers, once we snatch her up. PPK asked where is this bitch? Moufpiece said it is the GayMack goon in heels. PPK stopped and took a deep breath. Guess we should get ready for war in the streets, once we snatch Ms. Bobbi. Yep, but the sooner, we get the info the sooner, we can release and be done with it. Problem is how can we get Ms. Bobbi alone long enough to snatch. LabRat stated, like most of the people in the city, they all are going crazy over the "Lick'em".

LabRat told PPK, he didn't touch the stuff, because it is quite addictive. PPK stated he knew all about it and would make the necessary arrangements to set everything in motion.

Ms. Bobbi was on a mission of Zen, in other words SPA Day. GayMack was gone for the day to get some therapy treatments, from a new doctor that was recommended by Mynt. Ms. Bobbi was just happy and bouncing because a lot of their small problems had disappeared thanks to the Lick'em pills. She was able to setup a steady stream of drugs and users to get, majority of their street revenue back on point. Her biggest concern was getting rid of all the old baggage that Hunter left around. Ms. Bobbi decided not to allow business thoughts to ruin her spa day. Ms. Bobbi went through the kitchen entrance into the secure private VIP area. She gave the security guy her left wrist. He waved an ultraviolet light over it and the secret wall opened. Ms. Bobbi stepped inside and was immersed into pure sexual relaxation. The little host walked up to her and immediately began waving a wand in front of her. The device turned colors as it automatically put together a relaxation package for her. The host escorted Ms. Bobbi to her room and told her to fully undress and then step into the scan booth. From there the automated table will move her around to every area apart of her VIP package. Ms. Bobbi asked the host to forward all her messages to the automated receiver because she didn't want to be disturbed. Ms. Bobbi stripped and allowed herself to be immersed into the world of the autonomous massage. The robotics adjusted the table to fit Ms. Bobbi perfectly. Ms. Bobbi went into shutdown and just allowed herself to be relaxed.

Eow was still on a mission of finding Frequency. He had used all the small server system to try and get CMAX back online fully. Eow lost communications over the wide network due to the explosion at the Hangar. Finding LabRat and getting information from him was not easy. The kid was scared and now Frequency was missing, presumed dead. Eow dismissed the thought when he got a faint hit from one of the older secure bots that strengthen his faith, that the F.R.E.A.K.S. had taken a major hit, but were still in the fight. Eow was so intent on getting the network backup that he didn't pay attention to the camera feed from the automation robot. Frequency left facial recognition warnings, in all the camera systems just in case a major disaster happened. The software picked up on Ms. Bobbi from the body scan. Eow finished working on the communication problem, when the alert sounded a reminder. Eow looked and saw the face of pure evil and almost fell out of his chair trying to scramble. He was about to call Mack Master when he saw movement on top of the facility. Eow then recognized that the LMG was about to raid the building. Eow immediately set all cameras to motion memory. The place had hundreds of cameras and by allowing the computers to handle it, he would be able to see everything.

LMG stormed hard and fast, shooting and hitting anyone that moved. Ms. Bobbi was totally out of it and didn't feel a thing when the attacker stuck her with a needle full of tranquilizers. The men left Ms. Bobbi on the table and just covered her up. One of the guys was given orders to unplug the whole system and bring it with them. Ms. Bobbi and the robots were carted off as quickly as the attack happen. She was rolled into a truck and was taken to PPK. Eow immediately copied all the footage then doctored it so the police would not have an ideal of what was really going on.

PPK looked at Ms. Bobbi hanging from her ankles. He really enjoyed this robotic table, they had snatched from the spa. Thanks to LabRat and his computer skills, he had taken over control of the automation and was allowing PPK to use voice commands to torture Ms. Bobbi. As a hostage Ms. Bobbi mouth was getting her in a lot of trouble. The more PPK tortured her the more she talked trash and reminded him that she had bigger balls than he did. PPK didn't like not being able to kill this tranny, but he needed to know, who hurt his brother. PPK was about to chalk it up as a loss, when Moufpiece asked him, if she could give it a try. PPK nodded and then Moufpiece had the robots to hold Ms. Bobbi in the air spread eagle, facedown. She then had one of the smaller robotic arms inserted in Ms. Bobbi ass. Ms. Bobbi let out a yell of pleasure pain. PPK then told the software to vibrate in increments of 100 rpm. Ms. Bobbi screamed and struggled but was at the mercy of PPK. Ms. Bobbi was crazed in suffering. She shuddered hard as the vibration and pain sent her into one of her kill crazy orgasms. The robots drained her like a cow on a milk farm until she passed out. Ms. Bobbi was instantly hit with ice cold water that sent her into a cold-hearted scream. She was trying to break free when PPK told the software to increase the vibrations until it ripped its inside apart. Ms. Bobbi was screaming in pain. She lost all sanity and stated that GayMack and Mr. Mario was going to make them all pay for this shit. PPK increase the vibration but the circuitry in the building tripped off saving Ms. Bobbi from sure death. PPK phone rang and it was Eow, telling them to move quick, a helicopter was heading straight their way. PPK was about to leave and then Ms. Bobbi grabbed him by the pants leg. She told him he had fucked up and what Mario did to his brother would not be nothing compared to what she was going to do to him. PPK ran the power to full compacity, but shots fired into the building knocked the power out and saved Ms. Bobbi.

    PPK and his gang made a hastily retreat through the back of the building riding in ATV carts. The group of men came into the building and secured Ms. Bobbi 1st. They assessed the situation then brought in the GayMack. He was out of the wheel chair and now concerned for his right hand. She was battered, bruise, but alive. Ms. Bobbi tried to tell him what happen but GayMack stated that they had time to talk once she was healed and safe. They immediately boarded the helicopter and flew away. The pilot asked him did he want to torch the building and GayMack stated he just wanted to get as far as away from as possible, because the heat was about to swarm. Eow was already up to cleaning up behind this mess. He used the robots to roll themselves onto an old container and secure themselves in. He had one of their nanobirds go seal and lock the container to make it look like it was legit. Eow then hacked into the system and had a pickup schedule for it to go to another dock until he could find time to dissect it. Eow knew things were getting ready to explode and now, he had to keep an eye on LabRat, because he was in bad territory with no help. Eow knew he needed to get CMAX back online fully, but without Frequency or the Nanogem. The payback would be a longtime away. Plus, he had to explain to the rest of the freaks that they were about to go to war all over again.

## 14.) CarWreck

Webb was driving Impala SS like it was meant to be driven. He didn't give a damn about the consequences because it was going to be a bad day for someone. His baby girl called him crying and stated a police officer called her names and spit tobacco juice on her. Webb used the locator on his baby's cell to find her and he would show off, once he got there. By the time, he made it to the area he saw his little girl sitting on a bench outside of some food trucks. Webb pulled up and immediately knew he had been setup. The little girl was a doll and he saw the wiring connecting her to the bench. He immediately tried to contact CMAX but saw that it was no communication. By the time he looked up he saw the assault team on the move. He immediately took off in the big Chevy and would teach these idiots not to mess with a mechanic with a CTD "crash test dummy" tendencies. The big V8 roared as Webb punched it and spun the car around. He immediately headed for the bunker area where BDB was working on his truck. All the communication was down, more than likely these guys was using jamming device. Oh well, Webb decided to just tear up enough shit to make the authorities come running. He knew where all the local speed traps were and decided to see if the city finest had the good stuff under the hood. That was a good ideal until the automatic gun fire started. Webb laughed out loud because he just had the car sprayed with the nanopaint. The bullets just pinged off the car like pellets. Webb sent out a shout to Don1 and Frequency for convincing him to do it. Webb then headed for one of the worst paved streets in town. Driving it was like a mine field and navigating it at high speed could be disastrous. Webb pushed the big classic way passed 85 mph around corners and dodging through traffic. He passed a couple of squad cars pulled over writing tickets. He immediately put on breaks and spun the big Impala around. The assault team immediately fired shots but missed hitting anything important. Webb aimed the Impala directly for the 1st car. Incoming traffic was oblivious of what was going on until the 1st car swerved to miss Webb, hit a pothole and flip over the guard rail onto a field below. The officer tried to radio dispatch but a 2nd vehicle shot him and his car. The other officer called for backup, but soon met the fate of his comrade. Webb rammed the 2nd car from behind and pushed it into the path of a cement truck. The driver slammed on breaks, but he smashed into the car killing all inside. Webb moved around the car and headed straight for the bunker. He made it to the overpass when the hover drones came after him. Webb pushed the Impala hard. These drones were in pursuit and they were totally trying to kill him. Webb headed straight for the airport. He figured there had to be something around there that would jam the signal. Webb now wished he would have listened to Eow and put something in the car that could have shut down the drones, but he was being old fashion and thing pure V8 muscle and skillful driving would suffice. Webb got to a straight away that lead to a bridge under the Airport runway. The drones were following but not as precise. As soon as he got to the bridge he pulled over and popped the trunk. Webb pulled out a KSG shotgun with a laser targeting system. As soon as the drones pulled into range, he fired of two rounds and hit two of the drones that exploded on impact, taking out the rest of them. Now Webb was awestruck, because he didn't have any explosive rounds in the shotgun.

Webb jumped back into the car and headed for BDB. The communication system still was not connecting, and he could not make a call of any kind. Webb started to worry because that has not never happened, even when one of them was locked up. He immediately changed directions. Instead of going to a bunker he knew he had to ditch the Impala and move to something more lowkey. Webb was trying to come up with something, when his car was rammed from behind. Webb gunned the Impala and soon the chase was on again. This time he could see the drivers, it was 4 different vehicles all of them were driven by females. Webb laughed to himself thinking he was not going to be captured by a bunch of women in some toy cars. He stopped clowning and really started to drive the big Impala. He knew the crew didn't know their vehicles because they were trying to run him of the road with vehicle that didn't weigh enough. He quickly pushed one of the cars over enough so that he could squeeze by. He immediate got out in front of them when he heard the import coming up from behind extremely fast. He knew the cars had NOS systems setup on them so out running them was not an option. He looked up ahead and saw any fast-moving vehicles worst nightmare, a pothole.

    He immediately swerved over making the driver respond by switching lanes. The car hit the pothole and a tumbling it went and then exploded. Webb knew he couldn't out shoot or run them. He decided to stand his ground and have a little demolishing derby. He spun the car around headed up another street with the other chase cars in full pursuit. Next thing Webb sees is the local police and they are directing their pursuit to the other cars. Unlucky for them these cars have pursuit drones and they start to kamikaze the police cars as fast as they can show up. Webb see the traffic cameras and a news helicopter catching his every move, which means he soon should be able to get help from somebody. The females started firing high powered weapons at any officer that wanted to be a hero. They had one mission and that was to capture Webb and try not to kill him. Webb didn't understand what was happening, but he was not about to start asking question until he knew he was somewhere safe. Webb saw one of the assault team cars take out a police cruiser and was about to shoot the officer. He punched the big Impala and the ass end of the car swerved around knocking a car rim into the shooter. The shooter missed giving the officer time enough to pull her own weapon and killing the shooter.

Webb smiled at the officer then sped away with a custom Honda Ridgeline in full chase. Webb never liked that truck, it was nicknamed the Kaitlyn Jenner, of pickups. Webb knew he was getting low on fuel and figured that truck had to be caring all the communication jamming equipment. He pulled around a corner and aimed for a gas station near the area. He figured if he could crash the truck into the gas station that would clear up the jammed communications and alert everybody about what was going on. Webb pushed the button for the secondary fuel source to kick in and he headed straight for the gas station.

Ali had just finish getting the station pressure washed when he heard all the screeching tires and explosions. He tried to make a call to the authorities, but no communications were working. Ali saw the big black Impala headed straight for him with another vehicle in hot pursuit. Mr. Ali ran, screaming something about these crazy car racers. Webb saw the gas station a hoped Ali ran away from the station. He immediately served in front of the truck and made it swerve and hit the curb hard. The Ridgeline tumbled and hit the 1$^{st}$ set of gas pumps. The explosion was intense as the station was engulfed in flame. Webb saw Ali and told him to get into the car. Ali was screaming and shouting about his station when a bullet went through his shoulder knocking him into the car. Ali screamed as he fell through the open window and Webb lit the tires up and took off. The car got him down the street then spurted to a halt. Webb was out of gas and had nothing but a few shotgun rounds with him. Ali was screaming in pain and Webb went into the glove box and got out the 1st aid kit. He stabbed him with a morphine shot and told him he would be right back. Webb popped the trunk and loaded up everything he had with him. He was going to end this one way or the other. He pulled out the cellular tracker and put it in emergency mode. If any of the guys was around, they would find him. Webb counted 17 rounds and then he would have to make do with whatever he found to fight with. He ran away from the car up an embankment. He saw 3 women and a swarm of them drones, two of the women had assault weapons. Webb knew he had to take them out first. Problem was he had the wrong rounds for the job. Webb had to come up with something quick. He looked around the ground and saw the perfect trap. Webb took off down the hill and jumped over the guard rail. He landed firmly on a beaver dam. He took a gamble and fired a shot that exploded 4 of the drones. The rounds from the assault rifle hit close but Webb was ok. He started running down the side of the lake bed. One of the ladies started chasing him and stepped right on top of the beaver dam. Her foot went through the dam and she screamed in pain, the drone went over her head and

Webb fired a shot hitting it and making it explode killing her. Two rounds hit close and Webb return fire exploding two more of the drones and injuring the other lady with an assault rifle. Webb ran for the rifle but was cut off by one of the drones. Webb jumped onto a tree limb that accidently went into one of the drone's propeller blades. The drone hit the ground but didn't explode. Webb picked up a rock and broke the other blades and antenna. He then went to go get that assault rifle. He made it to the woman, who was struggling to get to her weapon. Webb stepped on her hand as soon as she was about to touch it. Webb grabbed the rifle and was about to ask the woman a question, when her head exploded from a gunshot. Webb ran as more shots peppered around him. Webb jumped and hid behind a downed tree. He then heard someone talking over a radio. Webb looked closer and it was the drone he disabled earlier. It was the communication piece he thought was in the truck he blew up. He heard the last woman talking to someone about how bad things had gone and only her and a few drones were left. The voice over the radio stated don't come back unless you fulfills the mission. Webb knew he was in the fight with the last man standing. He was not going to be planted today and that was the end of that thought.

Webb checked the rifle and his shotgun. He had more than enough to finish the job and still take out the drones. Webb grabbed the drone and powered it down. He then started looking for the woman, who commanded all his attention. Webb thought like any other person, he headed back to the car, because it made since. Just as he thought he heard the drones headed in the same direction. The woman had made it back to the street and was headed for an SUV parked not too far from the Impala. Webb knew the Impala would do him no good out of gas and was preparing himself for a fight to the death. Just about the time he had prepared himself, the black smoke of the big diesel truck showed up. It was BDB and he was wearing full nanotach gear. Webb knew the big man would handle the rest. As soon as he was about to jump up the drones swarmed him, and he started to take them out. Some them exploded others just dropped but they were coming out of the back of the SUV.

    BDB saw Webb battling and he immediately joined in. Webb told the big man he was glad to see him. BDB told Webb to rest and let him show him how it is done. BDB pulled out the new prototype minigun and let it loose. The swarm disappeared as fast as they could take flight. The leader saw that she was out gunned and immediately took off in the SUV. BDB and Webb ran back to the big diesel and immediately began a new chase, this time they were doing the chasing. BDB caught up to her and was about to ram her when more drones flew out of the back. BDB yelled hold on as the drones hit the big truck and exploded on impact. The woman laughed as she saw the black smoke of the explosion. She pulled out a cell and started to make a call. Suddenly, the bed of the big diesel crashed on top of the SUV. Webb and BDB sat in the Impala waiting on the authorities. Ali asked them how he would explain this to his insurance. Webb started to tell him everything that happen, but BDB cut him off and told him to keep is simple and say it was a CarWreck.

## 15.) Geeks & God

And let there be light, so the Bible said, but who wrote it in the Bible was the question in Frequency's mind. Frequency watched the show about the Naked Archeologist, and How he went out to seek the truth in the scriptures of old. Frequency mind weighed heavy in thought of how technology was making the writings of old into more of hearsay than fact. Frequency was raised to believe in God, but everything after that seemed illogical to him. There was not one biblical scholar, who could beat today's technology on the teachings of old. Frequency knew that if he became outspoken on this subject, he would have to deal with all the people, who were raising to believe in all the teachings of the Bible. Frequency heart weighed heavy just thinking about Doc, and how adamant she was about religion. Frequency knew that if he just embraced religion, he could probably have a love of life back in his world. That felt wrong no matter how he tried to view the situation in his mind. Frequency then thought about how the good Reverend Hippocrite used religion to make himself a very wealthy man and then turned to use it for wrong. Frequency knew he had to make an example of him and all others like him. Frequency then had CMAX put together a biblical time chart and then matched it against all other religious writing during that same period. Frequency made sure that CMAX used all the tribal data and Egyptian data gathered by archeologist to ensure that when he started going after the Reverend, he had some knowledge to back up his plan to throw the Reverend off balance in his attack towards him and his religion.

    The Reverend had been in contact with Jahoodi for years and he knew he would have the resources to get any adversary off his case. The

Reverend had locked down all the former trafficking groups and now was siphoning them for all their finances. The Reverend enjoyed the power of persuasion and all the inequities that came with it. The Reverend had been given some info that one of Hunter's people was alone in New York and was an easy target to snatch and use as leverage against Hunter. The Reverend gave the ok before discussing it with any of his partner in crime. The Reverend shutdown for the evening and went to enjoy himself in the wealth he had amassed.

The mood in the small surgery center was quite nullifying. A bus load of teens had a bad crash directly in front of the building. Everybody was scrambling trying to help all that was injured. Teez had just ended her shift when all the chaos unfolded. Teez took her rolling bag and put it in the nearest closet for safe keeping. She immediately started helping with injured. The duty nurse saw her and told her that she had already marked her off for the night. Teez told the Duty nurse that she could just use the entry camera footage for her time management and then adjust it later. The duty nurse gave her the thumbs up and they begin triaging patients. Teez was so caught up in the moment she didn't feel her cellphone. She went into her pocket to receive it and it wasn't there. She immediately went to the desk and called it. She heard "Let's go Crazy" playing in the hall. She looked up and saw one of the teens walking away with her phone. Teez made sure everybody had help and then she immediately went looking for this teen with her phone. Teez hurried around the corner and then she saw the elevator door close. It went up the crossover deck which led into the garage. Teez took the stairs and made it up to the crossover floor. She went down towards the ultraviolet light area and she saw the strangest thing. On the back of the neck of this teenager was a hidden barcode. When she stepped out of the light the code was not visible. By the time Teez made it to the door the girl got into a car and it sped off down toward the exit ramp. Teez was pissed, she really liked her phone. Teez then realized that she was now stuck in the garage and had to walk down to the main entrance. It was good that she had left all her identifications locked up inside of her rolling case. This could have been a long week without ID. Teez was about to get to the main floor when all the gunshots started. She immediately dropped down behind a car and started looking for a way out. She was freaking out because she couldn't open any of the doors without her work ID. Then she remembered about catching the interns taping the side door, so they could go out and vape. Teez moved quickly and quietly as she heard more men screaming and firing shots. Teez was afraid as she saw all the people being locked up in different rooms. Teez didn't know what was happening but she knew she didn't want any parts of it. Teez made it to the back side of the entry closet then went inside. She got to her case and struggled with the combination code. She tried to calm herself down, but she heard one of the guys stating that the number that they had was not in the building. Teez saw one of them grab the duty nurse and take her in the elevator. Teez saw that the men were heavily armed and well trained. Teez immediately opened her case and put on a different set of clothes. She figured they were looking for a hospital employee and she was not going to look anything like that. Teez got dressed and quietly slipped back outside of the center. She quickly stepped into the alley and immediately made it onto the sidewalk of busy walking New York traffic. Teez saw a courier van parked in front of the center and had to maintain her composure. The duty nurse had been thrown from the crossover and landed directly on top of it. People and police were scattering trying to figure out what was going on. Teez just walked away hurriedly not trying to draw attention to herself. She knew she couldn't go to her apartment because whoever was responsible for this, would go there looking for her.

Teez was scared and it was getting late. She had a lifeline to call but then that had another set of ramifications. Teez had some cash in her rolling case and went and found a room for the night. She paid for room service and then contemplated her next move. Teez tried to relax her mind so that she could come up with a way to get out the city without anyone noticing her. She heard a band playing some live music outside. It had a reggae vibe to it, but it was jazzy to hear ears. Teez knew it wasn't safe but something about the song that they were playing had her full attention. Teez walked outside and checked out her surroundings, nothing seemed out of place, just an old Jag with European tags on the front. Teez didn't think nothing of it, so she went inside the lounge and looked around. The band started to play the song again but this time they played it with a little more rock to it.

The crowd really liked it and so did Teez. She found her a seat in the back corner and ordered a drink. The waiter brought her some fish and chips with a glass of sweet tea. Teez was shocked because she didn't order any food or nonalcoholic beverage. The band started playing a song written by Sting and Teez laughed. She was about to get up when a gentleman, stepped to her table and spoke with a British accent. Teez interrupted him and said let me guess, an English in New York. The man said yes, my name is Xander Englishman. They both laughed as the band sang the song "An Englishman in New York. Xander asked if he could join her and she said ok. Teez asked why he was here in the lounge. X was told that a very important person was missing in the city and that she was to stubborn to call for help. Teez laughed out loud as she stated guilty as charged. X smiled and said good, now that he had found this person, the hard part was getting her out of the city without getting them both killed. Teez looked shocked, she didn't understand why she was being persecuted. X explained to her that the center she was working in was a part of a trafficking network. Someone exposed the center and a shutdown crew was sent to clean up the details. One employee escaped and now is being hunted. Teez thought about the girl, who stole her phone. She must've tried to contact someone, and all hell broke loose. Teez dropped her head and X told her to eat. They would have to get moving as soon as possible, if they wanted to get a jump on the traffickers. The band leader announced that they were taking a break and would put on some music till then. X grabbed Teez by the hand and gave her the keys to the Jag.

Teez looked at him in confusion as he told her to get in the car and meet him around the corner. He repeated himself to ensure she understood and then, she got up and headed for the car. As soon as she made it to the door some guys busted in with guns drawn and the music started playing Terminator X by Public Enemy. Teez saw X come out of his waist coat with something sharp and the 1$^{st}$ guy fell to the ground dead. The other guys ran directly pass Teez as she got into the Jag and followed instructions. X meanwhile had picked up the Deadman's gun and made quick work of the street trash. He went inside one of the guys pockets and found a burner phone with explicit messages what to do with him and the woman. X laughed collected their weapons and moved to the rendezvous point. As he got their he heard his car running just as smooth. X had Teez open the trunk and he put the weapons inside except for one of them. He escorted Teez to the passenger side and then had her show him to where she was staying. He told Teez to get her belongings and to move fast, they were headed back to her old home so that she would be safe. Teez didn't argue and hurriedly got her rolling case and went back to the car. X saw the rolling case and had an ideal why this woman was so important. X kept his thoughts to his self and the two headed out of the city.

Mynt got the message that his squad was wasted, and the Reverend must've notified Jahoodi about the center. They were all confused on

How one of the workers had escaped and was on the run. Mynt knew the authorities didn't have anybody in the area that was working undercover, because they had dirt or something on all current officials. They really needed to find Hunter because his carcass or something could clear the way for them to be unstoppable. Jahoodi would have another local squad check out things because he didn't trust none of them. Mynt knew Jahoodi was setting up a Somali hit squad to clean up all the loose ends and he didn't want no parts of that. It was bad enough the GayMack had setup another club and was back controlling the nightlife.

The Saggy Boi was the spot for anything goes night time action. No authorities would be caught dead anywhere near there because of the dirt Ms. Bobbi had on them. No was safe from their interactions. They were sexing any and everything, with nothing spared but the roaches. The place was a cornucopia of Sodom and Gomorrah, Caligula, and the film archives of unimaginable sexual desires filled to compacity. Within the realm of this area only the supplier of the drugs, the keeper of the data, and the purveyor of the property reigned supreme. They were referring to as Geeks & Gods.

# 16.) Junk N Trunk

The Saggy Boi grand opening was the talk of the town. The who's who, of the underground scene would be there. Everybody in the adult entertainment genre confirmed their VIP invitations and the word on the street that Ms. Bobbi was going to make an entrance. That alone signaled the authorities that their judicial system had been compromised dearly. The GayMack was finishing up some last-minute details as he exited his stateroom onboard his newly acquired yacht. Ms. Bobbi was back up moving around fully healed thanks to Dr. Stamps and Roz. They all had now formed an alliance thanks to Jahoodi and the Somali Swordsman pirates. These men had basically swarmed, maimed, and murdered every obstacle in their path except for the good reverend. Why Jahoodi would not attack him was a misery, but since Jahoodi had all the money and power, there was no need to go against that.

Moufpiece was in agony and she couldn't believe that she had gotten taken from PPK. The swarm of the Somali gang took out a lot of PPK's core group and sent him off into hiding. LabRat was nowhere to be found and that was a good thing, because she overhead one of the Somali's stating that he was going to be fucked and beheaded at the pleasure of Ms. Bobbi. Moufpiece had been put into a torture rack made for an S&M dominatrix. They had her legs locked behind her arms and she was naked. They device would move her around any position her suiter wanted her in for maximum penetration into any orifice with little lubrication. They were told not to touch her or else they would be given to Ms. Bobbi as toy to have her way with.

     Roz and Mynt was loving the life out on the FoF. No worries and the security of Jahoodi was awesome. Roz wondered what Frequency would say if he knew she was behind his demise. The thought made her laugh as she saw another one of the Gaymack's girls walking up to her. Roz was told that they had all the players contained and the grand opening was now in full effect. Roz then asked what the big deal and girl was stated that Ms. Bobbi wanted to make sure everyone was there, so that they can make the announcement that they were back in charge and all that had a problem would be severely dealt with. Roz nodded and handed the girl her empty glass. The girl pulled her apron to the side and pissed in the glass. Roz immediately jumped up because the girl was a transwoman. Roz was about to go off, when she heard, Jahoodi behind her state to the woman, if she wastes a drop of that piss on his chair, she would not see the hint of daylight ever again.

    The transwoman cupped the urine and ran away trying not to spill a drop. Roz turned to Jahoodi and he sincerely apologized for the insubordination. He assured Roz nothing of that nature would ever happen again. Jahoodi turned and walked away, he now had to teach all the new people that his words and rules were not to be messed with in any form or fashion. Jahoodi had L1 get his chambers ready because it was time to teach all the new inhabitants a lesson in discipline. L1 nodded and was about to leave, when Jahoodi told him to find the little prissy boi and make him comfortable in my chamber.

Jahoodi had every video onboard the FOF playing out his discipline of the misbehaved guest of the GayMack. He 1st used The Chair

of Torture ([Torture](#)) there are many variants of the chair. They all have one thing in common: spikes cover the back, arm-rests, seat, leg-rests and foot-rests. The number of spikes in one of these chairs' ranges from 500 to 1,500.

To avoid movement, the victim's wrists were tied to the chair or, in one version, two bars pushed the arms against arm-rests for the spikes to penetrate the flesh even further. In some versions, there were holes under the chair's bottom where the torturer placed coal to cause severe burns while the victim remained conscious.

He then explains to all the guest why this individual was singled out for Jahoodi entertainment. Jahoodi made sure the GayMack and all his guest understood that there was no tolerance for going against his wishes. He then used his 2nd method of discipline The Pear of Anguish ([Torture](#)).

The Pear of Anguish was used during the Middle Ages to torture women who conducted a miscarriage, liars, blasphemers and homosexuals. A pear-shaped instrument was inserted into one of the victim's orifices: the vagina for women, the anus for homosexuals and the mouth for liars and blasphemers. The instrument consisted of four leaves that slowly separated from each other as the torturer turned the screw at the top. It was the torturer's decision to simply tear the skin or expand the "pear" to its maximum and mutilate the victim. The Pear of Anguish was usually very adorned to differentiate between the anal, vaginal and oral pears. They also varied in size accordingly.

Jahoodi had his favorite music played as he commenced to administering the different means of torture. Jahoodi knew the limits to keep the victim alive until he was ready to end the life slowly. The transwoman was convulsing as the device stretched and popped her open. Jahoodi shut the machine down and slowly relieved the pressure on it to increase the pain and slowly rip apart the body from the inside. Jahoodi then decided it would be enough of that and he wanted to hear why this transwoman felt it was ok to relieve itself in one of his guest wine glass. So Jahoodi decided on the 3rd method the Judas Cradle ([Torture](Torture)) The Judas Cradle, a terrible medieval torture where the victim would be placed on top of a pyramid-like seat. The victim's feet were tied to each other in a way that moving one leg would force the other

to move as well - increasing pain. The triangular-shaped end of the judas cradle was inserted in the victim's anus or vagina. This torture could last, depending on anywhere from a few hours to complete days. Jahoodi asked the transwoman why she felt the need to show off the ignorance of manners on his premises. The transwoman tried to speak, but still had the pear of anguish stuck in her mouth. Jahoodi saw that and removed it slowly from her mouth as he increases the pressure of the Judas cradle. The transwoman was screaming in agony as soon as the pear left her lips. Jahoodi smiled at the amazing toughness of this individual. Jahoodi asked her again, who told you that it was ok. The transwoman didn't say a word as the cradle increase pressure and she passed out. Jahoodi had her pulled off the machine and place in the human vice for methods 4, 5, and 6.

Jahoodi set her up in #4 the

Head Crusher ([Torture](#))The head crusher was widely used during most of the Middle Ages, especially the [Inquisition](#). With the chin placed over the bottom bar and the head under the upper cap, the torturer slowly turned the screw pressing the bar against the cap.

This resulted in the head being slowly compressed. First the teeth are shattered into the jaw; then the victim slowly died with agonizing pain, but not before his eyes were squeezed from his sockets.

This instrument was a formidable way to extract confessions from victims as the period of pain could be prolonged for many hours if the torturer chose to. This could be done by repeatedly turning the screw both ways. If the torture was stopped midway, the victim often had irreparable damage done to the brain, jaw or eyes. Many variants of this instrument existed, some that had small containers in front of the eyes to receive them as they fell out of their sockets. #5 The Spanish

Spider (Torture) The Spider was designed to mutilate a woman's breasts. The spider is nothing but a variation of the famous [breast ripper](), though the spider is believed to be more painful. The spider was usually chained to the wall - the claws were usually heated before being fixed to the woman's breasts. The torturer just had to pull the woman away from the wall to successfully remove her breasts. This was a brutal punishment that often resulted in the victim's demise. Fortunately, it was rarely used and only reserved for women who committed very serious crimes:

deliberate miscarriage and adultery. And last #6, Crocodile Shears

(Torture) , The crocodile shears interior design closely resembles a tube containing numerous spikes on both ends. Although it was sometimes used to mutilate the fingers, its most common purpose was to mutilate a man's penis. This torture method was frequently also a way to execute victims, as the arterial damage that ensued from the torture provoked death very frequently.

As the transwoman was revived, she now understood that she was about to be ripped apart. Jahoodi explained each of the six methods of discipline he was using on her and explained to her that he would allow her to live, if she gave him a reason why she chose to disobey him. The transwoman then told Jahoodi could he put the pear back in her ass, because she was about to get a nutt. Jahoodi increase the pressure to a higher threshold which simultaneously, ripped her implants off at the nipples, shredded her penis, and crushed her head with body fluids squirting all over the place. Jahoodi started dancing and celebrating as he took another victims life and decided to attend the clubs grand opening in celebration.

Mr. Mario and Ms. Bobbi were stunned but didn't say a word. They knew they had fucked up but was not about to say a word to no one. Ms. Bobbi hurriedly got showered and dress for her grand entrance tonight. Mr. Mario knew he had to watch his back, because he was not trying to get on Jahoodi torture chamber list.

The GayMack had already made the call that he needed to make. He wanted the good reverend to confirm that he would be attending tonight's festivities. The Rev assured him that he would be there in the flesh, no if and buts about it.

The lightshow and entrances were made like an old school players ball. The infamous showed up and out making sure they were seen and heard throughout the facility. All the players showed driving their best cars and security was on point. Every player had his own crew setup, just in case, anything popped off they could get out and no loss of any of their people. The discipline of the transwoman was shown to everybody in attendance so that they understood that no fighting amongst them would be tolerated.

The show began, and Ms. Bobbi made her grand entrance the rest of the players were in awe as Ms.

Bobbi was looking like not one hair had been touch on her head, even though the streets had her as half dead. The party began and the "Lick-Em" batches were dispersed out to all the players. Roz and Mynt had made a new batch derived from the old batch and an LSD mixture. Everybody went wild as the music pumped and the fog machine pumped out the new mist batch for all the non-users in the crowd. Reverend Hippocrite had gotten word of this batch and made sure he had already injected himself and his entire crew with it. Rev was thankful his new friend Lady Madame informed him of what was really going to happen. The good reverend decided to make one of his deaconesses a sexual Guinea pig and use her to pass on the pleasure to his most doubtful parishioners. It was so funny to present them with blackmail footage of them in coitus with children, other parishioners, and some animals. Reverend Hippocrite had all the financial data entitlements to afford him a very long and lavish lifestyle. His biggest concern was why did GayMack want him here inside of his den of inequity. The club roof opened and suddenly, a man and woman seemed to be descending from the sky. Reverend knew he wasn't high, so he knew this had to be some powerful people. Jahoodi and Tavia decided to drone descend their way into the party. To his surprise Ms. Bobbi didn't have a word to say, but that may have been because she had a mouthful of some young stud penis. Jahoodi just ignored them all and went to have a discussion with the main reason he was there.

Jahoodi, GayMack, and the Reverend went into a room and started talking about what they had to do to keep the peace, since Frequency and his crew were defeated. Jahoodi was quick to mention that until a body was found, they were assuming, he was beaten badly but not dead yet. Jahoodi then told them that he had the reactor that he needed for another project, but unfortunately was lost. Thanks to his mate Tavia they were able to directly tie the reactor into one of the areas fault plates. Ensuring no interference from the authorities, or they would be causing a shift in the plates that would cause a cataclysmic event that would destroy the entire city. The Reverend now understood why the Russian woman was there. Jahoodi was playing a game of word domination. The Reverend decided to stay in agreements with Jahoodi and stated that he needed reassurance that he had the security to keep him and his flock safe. At that time L1 walked in and stated it was time for the announcement. Everyone stood and followed them out to the main stage. Jahoodi took the mic and said to everyone, let me have your attention. Everyone stopped and looked up at Jahoodi with pure submission. The new batch of mist had a special blend that gave Jahoodi full mind control. Jahoodi told everyone that he was now in charge of the underground and that GayMack and Reverend Hipp, was his right and left hand. He stated as of this day forward all work goes through him and no other. And to celebrate this tremendous occasion, he was giving away one of Frequency Hunter's crew to the highest bidder. The curtain was pulled off Moufpiece trapped in the domintrix chamber oiled down and fully naked. The auctioneer will start the bidding and it is a no minimum bid, cash or equivalent, to win!!!

Xander Englishman and Teez was at the all-night café waiting on the next set of instructions of where to go. The old Jag got them there, but it was showing signs of some maintenance. They could hear and see some of the displays from the building down the street from them but didn't have a clue what was going on. Teez ordered some food and coffee and Xander went to the men's room. As he was on the stool he overheard outside of the café, the employees were talking about all the events happing in the city. They were tripping how some old school players came back into the city and was now running the underground. They were chatting it up when the cook told them to come back inside because the order was ready, and he wasn't giving discount because of cold food. Xander cleaned up and allowed time for Teez to go to the restroom to clean up and come back and enjoy her food. Teez had no idea what was going on and if it had anything to do with the attacks. She hurriedly finished her meal and coffee and headed back to the Jag to go find her peoples. Xander paid for the meal and went outside to get into the car, but it wasn't there, and neither was Teez. Xander looked around and saw the car had been moved to a dark alley and Teez was inside. Xander went to the car and asked Teez what was going on, and she pointed up overhead. The two watched as two people floated over the top of the café and down toward the club. The two of them remained silent as the heavy armored vehicle followed down the road with a gang of men in tow. Xander knew this was not good because none of these people were the authorities and they were cornered with no way out now. Xander sat in the car and watched the movement of all the calamity starting to settle. He knew if he cranked the jag all hell would take place. The two had to wait it out and hope for an opening to escape. Then out of nowhere the Righteous One caravan came down the street and headed for the club. This was possibly the break they needed until Teez saw Madame. Teez immediately tried to go after her but Xander grabbed her and would not let her go. Teez now knew that these people had something to do with all the craziness that was happening. Teez remembered everything that Madame and her gang did to her and her family. If it wasn't for Frequency, she would have been wiped out with the rest of them. Teez knew she had to get to a phone and contact Mack Master, if anyone knew what was going on, he would. As the last of the cars went by something in the distance caught their attention. This one guy was walking up to the different whores, trannies, street people and players asking for directions. The guy was pushed around by one of the Somali's but then he knocked him out with one punch. The guy disappeared into the alleyways before all the other guys ran after him. Xander and Teez immediately jumped at the chance and got the hell out of there. Xander made sure no one followed them as they made their escape. As soon as they saw a police substation Teez made Xander pull over and let her go and try to contact Mack Master. Teez remembered what she was told from day one. Never assume all the officers you talk to are there to help you. So, she went up to the main desk and asked, where do you pay your traffic fines at. The officer at the desk told Teez that the automated system was over there, but it was out of service and she would have to use the drop envelope payment. That was perfect for what Teez needed, she went to the drop box, filled out an envelope and then left. She hurriedly got into the Jag and they sped off to a spot where Mack Master, would know where to go. Next Teez told Xander that his connect would have to deal with Mack Master, if they had a problem with the authorities, because what she had just seen was not good for the home team. As soon as they got to the old garage Teez knew things had really gone bad. It was totally abandoned. Teez went into the old office and removed the file cabinet off the false floor. There it was like they said it always would be. Teez grabbed the bag from the floor, when a vehicle pulled up and a familiar voice called out. Teez stepped into the light and there was Eow and Mack Master. A sight for sore eyes. Eow then turned to Xander and stated that he could come with them if she still had the carrying case.

Xander got the case from the trunk of the jag and got into the vehicle and handed it to EoW. Teez was confused then it all made sense, when she remembered that Frequency always had someone close to her. Teez held her emotions till they were out of the city and had a chance to find out what was really going on, but until then she now knew she was going to be ok, hopefully.

Madame was over excited that she was able to get back in the city and get her money back on point quickly. The good Reverend was quick to help her get the finances up to par thanks to the simple people of his flock. The one thing she was not going to do was let her guard down on, the fact that her husband was alive and hellbent on revenge, after she turned her back on him and poison him. Oh, well that was a thought for later as the celebration was happening in the newest underground spot in town, the Saggy Boi. Madame saw that the good

Reverend had finished with his meeting and it was time to show him what happens when Madame makes up her mind to mesmerize you. Madame went up to the Reverend and lead him to one of the corridors that had a private room, she immediately went to work on the Reverend ensuring he would be coming more and more. The Reverend was lost in his pleasure and not paying attention to the buildup of smog outside of the room they were in. The pleasure crazed couple was deep into their sodomistic romp, when they

heard multiple explosions and shots going off.

Madame had just finished tying the Reverend up when they heard the authorities outside trying to raid the club. The two were amused by the efforts and Madame started back pleasuring the Reverend. Baton had somehow gotten and invitation to go inside of the club. Since she was on the run it was like someone wanted her there. As she was having a drink and trying to stay away from all the craziness going on, she saw how they had the young girl up on display for auction. This irritated her to no end, but now she had no one to help her, or that is what she thought. Baton walked around the club and was suddenly snatched from behind into a room and a very sharp knife was to her neck. The voice asked her a specific question concerning Madame and the Reverend. Baton told the individual everything that had transpired, up to the point of Madame trying to kill her husband for the sake of greed and power. Baton even explained how Madame turned the tables on her and had her entire family killed because she didn't want her to tell the truth. The voice then instructed Baton to leave this place immediately. Baton said ok then she was released. She immediately tried to leave but one of the high participants decided he wanted some of that woman walking by and no was not the answer. Baton knew if she put up a fight, she would be ganged up on and that was not a good thing around a bunch of nude high men. She saw her way out by picking up an empty bottle of oil and stating that she would take care of him as soon as she got more. The man eyes were glazed over as another dancer grabbed him and started pleasing him right then and there. Baton moved fast to the kitchen area and then back down toward the garage area to make her escape. Mynt and Roz were in the garage preparing to set off the next batch of gas and virus, when they were told to get back to the FoF immediately. The club was about to be raided by the feds and Jahoodi didn't want any of his core people compromised. The two immediately set the timer on the fog machine and the multiple drums of TPV/PFV, then headed for the docks.

Whiteface overheard the whole conversation, but he didn't give a damn. He wanted everyone affiliated with his brother's killer's dead. Before he could get up into the building, all hell broke loose. PPK and some of the other factions started to raid the club. Madame and the Reverend came out of one of the rooms covered in oil. She was wearing nothing but a strap on and some electric nipple rings. Whiteface started firing off harpoons trying to kill and maimed everything in his path. Madame saw the movement of the individual in all white and instantly knew it was her time to leave. Madame grabbed her cash cow the good Reverend and headed for the automobiles. One of Whiteface's harpoons stuck her into a door by the dildo of the strap on that she was wearing. Reverend quickly unfastened her, and the two nude individuals made way to the garage area and escaped from the massacre. Whiteface was stabbing everything that moved around him and he somehow allowed Madame to get away. Meanwhile, PPK and what was left of his crew was battling to free Moufpiece. All the main players had made their escape by the time the feds started to arrive. Once they got into position to move in on the club, the only people that were left was dead, maimed, hungover or still trying to hump themselves to death. Layrock was pissed because he knew he would not be able to get back control of a bad deteriorating situation for everyone. And then as if that was not enough, the club erupted in an explosion, when one of the officers open a room with some junk in a trunk.

Saggy Boi Club

## 17.) Correbous

EoW had to make the call that he had only heard about a few times. Since CMAX was still not back to full operating order. The crew had been using all the old tactics to keep them alive until they could find their leader. EoW opened his first computer case and pulled out the 1$^{st}$ gen Nokia phone. As soon he put the battery in and powered up the phone started ringing. The voice on the other end asked for the code. EoW played back the recording between him and Doorshaker. The voice on the phone call told EoW to gather the entire crew at the designated area and make sure they understood that the time had come to retaliate.

The guys met up at the designated drop zone and Doorshaker had already parked the trailer and was awaiting payment. The reset of the crew showed up and immediately began cutting an opening in the top of the pipe. As they cut small pieces to ensure they could hold the pieces with the human power that they had available. They were able to get enough of an opening to see the ominous looking liquid inside. Immediately, LabRat told them that the glass enclosure that Frequency was standing in when the explosion happened was filled with this liquid. EoW knew he needed Don1 to restore the communication and Nano symmetry. As if the universe heard the thoughts in his mind Don1 flew in and landed outside of the area. He immediately ran in and knew what to do to get the active Nano particles back in sync with the computer algorithms. It didn't take long and then something amazing happened. The clean house robots converged onto the trailer and started setting up small inner network to move data. Next a Tesla autonomous rig pulled inside and then out of the trailer came the surgical robots. Somehow the artificial intelligence mode tapped into the backup logs of CMAX and they started restoring connections. Don1 knew this would take time and he had to tell the group that they would have to prepare themselves for the big attack, since now they knew, who was responsible for all these attacks. Eow received a call from Mack Master and he was told to meet him outside, they had to go get the rest of the items needed to get CMAX fully functionable.

Mack Master and Eow had Teez and Xander in the car and it didn't take them long to get them into the containment area with the rest of the crew. As soon as LabRat saw Teez he knew, it would not be long before Frequency would show up. Teez gave Don1 and all the crew a big hug for taking care of her. Now why wasn't Frequency here to greet her. Teez then called out to CMAX but there was no acknowledgement. Teez fell to her knees in tears because she knew things were very bad. Eow ran to her and picked her up off the floor and told her they needed for her to hold it together until they could figure out where Frequency is. About that time the rest of the crew started to gather around her and Don1 asked Teez where was the Nanogem. "Nanogem" is this what this is all about, Teez looked confused and upset. So, she asked Don1 to show her the film from the explosion of the hangar. By the time they had everything setup, BDB,

Webb, and Thunderhead Hawkins arrived. BDB was furious and ready to destroy something. Teez remembered him and asked BDB, how did Frequency know where she was. BDB said "CAMSHAFT" suddenly, a little droid popped out with a Gatling gun ready for war. Teez went into a corner to hide and the little droid stood in front of her not letting anyone get near her. Teez was amazed at how the droid responded to her concerns only. She knew he was spying on her but not with her own rolling bag. Teez hugged Camshaft and thank the little machine for being there for her.

Teez now understood how and why she was able to get out of some crazy situations. Now she had to find her man Frequency and the only clue to his whereabouts was this video. Teez asked the guys was there an area where she could get cleaned up and get started on the task at hand. Eow pointed to the private bath in the corner and she went inside to get cleaned up. Teez didn't feel right about the whole situation but she took a long hot shower and just let the water relax her aches and pain. Teez smiled to herself because she knew camshaft was at the door and no one could come in on her. She opened the bag that she had hidden in the old shop and her jewelry was still intact inside of the vacuumed sealed container. She hurriedly put on her jewelry and did a quick pose to see the view that Frequency fell in love with. She couldn't believe her piercing were still open. Teez heard Don1 scream out there he is and immediately she threw on some scrubs and ran barefooted out to the video area. Camshaft was right there with her making sure no one would get close to Teez. Don1 slowed the video down and did a directed laser image of the video which showed that Frequency was trapped in the section of the pipe that was in the containment area. Everyone converged to the pipe trying to see anything inside of the murky fluid. LabRat stated that someone had to get inside and see if they could find

Frequency. By the time they could try and stop her Teez jumped into the murky water looking for Frequency. Everyone was scrambling trying to throw something into the water to save Teez, but nothing happened. The water started bubbling and one of the guys came up with Teez scrubs ribbed into pieces. Something was happening, but they didn't understand what was happening.

It seemed like an eternity to Frequency, he was stuck in trance and he couldn't break free of it. It was like the stories of old when you had one of those paralyzed nightmares and your mother use to say that the Witch was riding you. Whatever it was, must've been massive because all his backup systems have not pulled him from this containment. Frequency tried to scream out, but his body would not respond. He tried kicking and punching but there was no movement. His only glimmer of hope was when he bounced off something and he was able to hit the letter E, the numbers 1 and 8. He didn't know who would receive it, but if it got to one of the guys, they could figure what to do next. That thought alone made him smile at the thought of the beautiful woman that had the secret to the crew. Frequency was so lost in the thought that he didn't realize another body touched him.

Frequency was able to open his eyes and there she was. The hook of one of the poles had ripped her clothes so she was totally naked. Teez grabbed him and immediately kissed him releasing the Nanogem, that were hidden inside of her body jewelry. The murky water started to swirl and suddenly Camshaft dropped an electrode into the water, which sent the fellas running for cover. The next thing they saw was the water clear up and a naked woman getting out of the tank. LabRat hurriedly grabbed a sheet and covered Teez but something was totally different about her. Her hair and eyes were lavender in color. She was slightly winded, but she didn't want to leave until everyone saw what she had seen. BDB knew what he needed to do so he went and turned full power on to the containment area. The communication network started humming then suddenly they heard "CMAX Online!! Everyone clapped for joy because that was the sign that Frequency Hunter was now back where he belonged. Everybody was so happy and then they saw something moving in the tank. The silhouette of a man stepped out of the tank and covered himself with a sheet. Nobody moved except Thunderhead Hawkins he was about to hit it with a shovel, when suddenly

CMAX stated that Frequency Hunter is alive and well, he should be restored to normal viewing in 10 seconds. And just as he stated the incredible happened, the crew had found Frequency Hunter.

Teez awoke in her bed that was put together for her comfort. She had been by Frequency's side since they resurrected him from the bottom of the tank, because he had been in and out of consciousness, since they found him. Teez had so many questions and nothing would stop her from getting even with Madame & Roz, because she knew they had something to do with this. Teez hurriedly showered and gotten dress. She noticed that the crystal pieces from her jewelry were missing and thought maybe it came out when she jumped into the tank for Frequency. She decided to replace it later, because now her focus was to get Frequency back to his old self. Teez went into the room and the nurse bot was sitting him up and Frequency's skin color was changing to the point that he blended in perfectly to his surroundings. He was having trouble talking so he had the robot bring him a keyboard. Frequency immediately had CMAX link in his brainwaves to the main computer's virtual dialogue commands. He linked in his new technoid DNA into CMAX then he was able to speak. Once he was able to speak, he started giving commands to figure out a way to extract the Nano camo serum out of him and duplicate it so that they could use it as a weapon against Jahoodi. Frequency was so into his vendetta, he didn't notice Teez in the room and she made sure he paid for it. Teez grabbed him by the balls and stated, I knew that Roz bitch would be the death of you. Frequency knew there was only one woman, who could get that close to him without any alerts. Frequency grabbed her wrist pulled her over his lap and exposed her naked bottom, he was about to spank her hard, but he noticed immediately the crystal from her jewelry was missing. Frequency hurriedly stood Teez up and ripped her shirt open. Both nipple piercings were missing there crystals also. Frequency now knew what saved him and Teez had no clue she was it. Frequency looked at Teez and kissed her deeply. Teez was ready for a fight. She had walked away from Frequency once before and she thought that they would never be together again. But here she was trying to pull away from him and nothing her mind thought would work did. Her body was in pure automatic get your spontaneous freak on. Teez couldn't believe how fast Frequency had her orgasming over and over. She knew the crew could hear her moaning and screaming as Frequency did not leave a pleasure spot on her body untouched. Teez knew she would not be the same after this, so she made it up in her mind that this would be the last man to love her and nothing, would get in between her and Frequency.

Frequency finally understood, what EoW was saying all the time. Nothing like sexing until you pass out. Frequency looked over at Teez's naked body and remembered how beautiful it was. He knew he better get up and get to work or they would be back at it again trying to see who would tap and say I am sorry $1^{st}$.

Frequency went to take a shower and the steam from the shower showed him a new feature that he could use to his advantage. Teez walked into the shower and screamed, Frequency grabbed her by the mouth and made her promise not to tell anyone what she saw. Teez immediately stated "Nitakungodea Milele" which stands for "I will be there 4 you" Frequency Hunter.

Frequency Hunter wanted to start his next transportation project. But instead He would dedicate the strategy to take out Jahoodi after the loved Lamborghini Frequency always wanted. The plan would have six

different attack modes named after a famous six horn bull called
# "Correbous".

## 18.) Done Deal

Jahoodi declared War on anything Frequency and his squads loved. He dispatched his assassins with strict instruction to make a death scene for the history books. Jahoodi also decided to personally snatch someone from Hunter's crew and make a meal out of them for his favorite pet. Jahoodi traveled throughout the city in a one-off Genesis automobile, he had everything at the ready to get him and his captive out to his private island, where he would disappear back into the sands of time.


Toothpick Fish Virus (TPV) and Blow Fish Virus (BFV) were all deployed. The TPV/BFV were placed into all the city's water source. The rest was for Jahoodi and his mercenaries to use in battle. E18 had linked everything into CMAX and the machine code did the rest. Everybody had been put on alert and warned that if they see Jahoodi just post a picture anywhere on the internet. The rest would be handled by the F.R.E.A.K.S.

Mack Master and all the police force was on high alert after all the attacks. The last thing they needed was an all-out war, but this was totally out of their hands and into the hands of a mad man. Thunderhead Hawkins was glad to see his boys, but he was concerned they was a little out gunned and outnumbered. Both men assured their pops that they were fully briefed and had brought the right tools for the job. Cutter Hawkins brought a (Dillion Aero The Ammunition Counter is a one-piece clip-on design that works with Standard Armament or Nobles feed chutes using 7.62mm NATO ammunition with M-13 links minigun) and Chopper Hawkins brought a (The

GECAL 50, officially designated by the United States military as the GAU-19/A, is an electrically driven Gatling gun that fires the .50 BMG (12.7×99mm) cartridge.). E18 made sure everyone could communicate accordingly and go get in their designated area and be prepared to fully engage whatever they came with.

    Jahoodi was contacted by Roz and she gave him the news that Hunter was alive and now he was going to be looking for him. Jahoodi asked her could she verify the news, and Roz told Jahoodi look at the movement of the hood boys. They have disappeared like something is about to come gunning and they don't want no parts of it. Jahoodi alerted the Somali gang and told them of the news. L1 smiled and screamed out it's time for the hyenas to feed. The assassins had already started their reign of murder and mayhem. They started taking out all Jahoodi's direct connects in the city and then they started preying on the LGBTQ communities by injecting them with the TFV/BFV. They had a very distinctive calling card that would have people all over the city hiding in shear fear. Meanwhile, Jahoodi had found the perfect person to bring Hunter to his very doorstep.

Webb was just getting comfortable in his witness protection hideout, when all hell broke loose. Men with machine guns started taking out his protection like they were paper targets. Webb saw a gun from one of the guards and he picked it up and started to put it to use. The men started throwing tear gas into the area and Webb saw the opportunity to escape. He ran through the door and was stuck with something by a woman that knew his name. Webb grabbed her mask and noticed that it was Roz. He tried to choke her but whatever she shot him up with took him out.

Jahoodi received the call that he was waiting for, so he and his squad headed for the connection point. Roz and Mynt knew they had just guaranteed themselves a very healthy payday and a guaranteed life away from all the chaos they had caused. Roz was so sure of herself that she made the one mistake that alerted Hunter's crew that they had just snatched Webb. Roz cut the tracking device off Webb's ankle that instantly sent out an alert.

EoW received the signal and immediately hacked every video feed in the area. One of the Somalis had a distinctive look about him and the system was able to track them. They all were riding in Genesis automobiles which also was a telltale sign and now Eow transferred all the information into a fully operational CMAX superstructure. Frequency was already enroute not waiting on anyone for backup. The Wazuma V8M moved like a cheetah possessed by Beelzebub himself. All the E18 crew was on high alert and joined Frequency on the chase to get Webb back. Roz didn't realize how big a mistake she made until they arrived to meet Jahoodi and a barrage of gun fire erupted. Jahoodi had Webb loaded onto the boat and immediately took off. Frequency made it to the dock just in time to see Roz with Jahoodi flipping him off.

Frequency Hunter hesitated then he saw that E18 had him covered and the war would become a naval battle. The Payara, Great White, Tiger fish, and Piranha were the sea breacher designs deployed and were perfect for Jahoodi and his mercenaries. Frequency Hunter ran and jumped into the air but was intercepted by his sea breacher designed water craft. These machines were fast and designed for the water kill. Each watercraft had been enhanced by E18 to triple the specifications of performance set by

the manufacturer. Everything that had a code designed by E18 was totally enhanced by the superstructure CMAX.

All the breachers were coded for the kill and the carnage left in their wake would be food for the fishes.

Jahoodi had thought he had time to gloat but the gunfire from the gun port caused alarm and panic for the Somali guard.

The Hawkins boys were wreaking havoc on the pirates, they were knocking off speed boats like they were in an arcade game. For every man that survived the sinking of his boat, there was the Great White sea breacher to end his days as a pirate. The Great White sea breacher had an internal tree type shredder that was meant for pure carnage of anything that it consumed. It was initially designed to clean waste and dead carcasses from the water, but now it was just a means to put an end to the terrorism of the Somali pirates once and for all. Each sea breacher was painted to look like the fish it was named after and they all had killer characteristics just like the animal it was named after.

Jahoodi and his crew made it to the FOF. He told L1 to destroy everything and leave no human alive to tell. L1 acknowledged Jahoodi and he went to the smaller helicopter to clean up the mess that they did not expect.

The FOF was on high alert when Jahoodi tied Webb to one of the poles in plain sight for anyone to see. Webb was so drugged he just slumped over and had no knowledge of what was happening. Jahoodi had all his guest meet him in his meeting room, where he had setup a torturous trap for everyone.

Jahoodi held his hand high holding the remote. He spoke to all his captured guest so that they could understand that their usefulness had come to an end. Mynt, Roz, Baton, Reverend Hippocrite, Madame, Quarter, Nickel, Dime, Mario, GayMack, and Ms. Bobbi, all were fastened to a chair and was awaiting their fate. No of them imagined that Jahoodi would go back on his word, but then they didn't truly no him. Jahoodi explained that he was pressed for time and thanks to one of the associates, the shit done got serious sooner than expected. Jahoodi went over to Roz and stated that her old acquaintances would be there shortly in time to soak up some of that reactor poison he was about to release. Ms. Bobbi immediately started squirming, she was not about to be reactor Guinea pig. As she struggled, against her strains Jahoodi walked over to her and loosened her locks. Ms. Bobbi was confused but she hesitantly stood up. Jahoodi asked, Ms. Bobbi to release the rest of the group by pressing the button on the remote that he placed on the table. Like a fool, she did and several of the associate were instantly dropped into the water below. Mario, Roz, and Ms. Bobbi survived. Ms. Bobbi instantly took off running and Jahoodi was in pursuit. Mario pushed back on his chair and it tripped Jahoodi up making him fall. Ms. Bobbi made it to another room where she was able to find the release mechanism to the chairs. Mario and Roz took off in different directions trying to save themselves. Jahoodi got up from the floor and gathered his thoughts. He was going for his guns when he heard the strangest sound over the FoF.

Don1 had followed Frequency and had rallied all the troops. Layrock was finally released from the hospital and now he was ready to kick some ass. All the crew was in the battle and ready to shut down this menace. Frequency used Don1 as a distraction to get onboard the FoF and rescue Webb. As soon as he got to his friend, he cut him loose from the pole and then administered a shockbot. Webb came alert quickly and was now ready for revenge. Jahoodi saw all of this from the main observation deck and he was pissed. He immediately started firing shots at the two men on the deck. Immediately, Webb and Frequency ran for cover. The seabreacher landed on the island and slid right in front of the two. Teez opened the canopy and handed the men Kriss assault weapons. Webb greeted Teez with a hearty hug. About that time Teez saw one of the people she been wanting to whoop since she been involved. Roz was trying to escape when suddenly, she was hit hard and thrown to the floor. She shook the blow off and look to see that it was Teez that hit her. Roz ran straight for Teez and the two women fell into a room that was unlocked. These two went at each other like MMA fighters driven by pure hatred for the other. Roz screamed at Teez every time she was able to get a punch or hair pull in. Teez would grab a body part and just twist it to hear Roz scream in pain. Teez caught Roz with a knee to the chin that sent her back peddling to the floor. Teez ran and soccer kicked Roz in the midsection knocking all the fight out of her. Teez was about to finish her, but suddenly another person tackled her. She was wet and smelled like a wet puppy. Madame was able to save herself after she was dumped into the sea. She climbed onboard to try and get a boat to get away from the chaos. Luckily, she found these two fighting and wanted to get involved. Teez quickly recovered and started throwing hands on this woman. Madame let Teez get a look at her face and then the fight was on. Teez was tired, but she was about to wreak havoc on both women. Teez put on her knuckles and punched Madame in the chin. Madame landed on the floor next to Roz. She was out for the count. Teez saw that the room had an external locking system, so she left both ladies locked inside to their own fate. Teez had to find the controls to the seismic plate reactors or all of this would be a waste. She used the navigation droid to take her directly to the area she needed to be in. Teez quickly connected the droid in giving

Don1 and Eow complete control of the FoF. The door slammed and Teez looked up to see Tavia with eyepiece on. Tavia was not expecting to see Teez and then realized what the problem was. Tavia threw the eyepiece at her and tried to run. Teez caught the device sat it on the desk then ran after Tavia.

Frequency was rushing around the FoF looking for Teez they had got separated and he did not need for anything to happen to her. Frequency would fire shots from the Kriss only when it was necessary, his main objective was to get off the island in one piece. Jahoodi swung the blade with all that he could muster but Frequency blocked it with the Kriss. The gun was knocked free, but Frequency caught Jahoodi square in the jaw, knocking him to the floor. Jahoodi scrambled to get his sword and then stood to use it on Frequency Hunter. Frequency was not in the mood to be sliced and diced so he quickly ran out of the room with Jahoodi in close pursuit. Frequency call out to EoW and he gave him instructions on how to get off the island. Frequency was almost free, when a hail of gun fire from a helicopter sent him into the torture room of Jahoodi. Frequency was trapped, and he needed to find a weapon. He went to the glass door and kicked it hard as he could shattering glass and giving him a fighting chance.

Jahoodi walked into the chamber cocky and braggadocios. He told Frequency that he would enjoy killing him nice and slow. Frequency found a shard of glass shaped like the state of Tennessee. He wrapped it in a strip of curtain he tore from the wall. Frequency stepped into the clearing to see Jahoodi with the sword. Jahoodi laughed and stated you chose the wrong weapon for the job at hand. Jahoodi charged Frequency swing wildly Frequency waited for Jahoodi to make a mistake and then he made him pay dearly. The glass shard caught him directly in his shoulder. Jahoodi quickly dropped the sword and was hit with a barrage of punches and kicks. Frequency was not about to let his man live after all that he had caused. Jahoodi threw something into Frequency face and escaped out of the room. Frequency was blinded and needed help, he staggered rubbing his eyes trying to fix his vision. He was about to take another step and then he heard Teez scream stop. Frequency froze in his tracks and didn't move a muscle. Teez pulled him back from the opening and help clear his eyes. The crew had gotten what they wanted and now it was time to leave. Layrock had all the people he needed in custody and the rest was either dead or on the run. Jahoodi, Tavia and L1 had managed to escape.

All the water vehicles were still able to get them back to shore. Mack Master had contacted the military about Jahoodi FoF. They boarded the Airfish and headed back to land except for Frequency and Teez. They decided to take one of the seas breachers and enjoy some alone time.

Jahoodi, Tavia, and L1 sat in the small submersible waiting for everyone to leave. They figured a few guards would be left behind but they would release a shockwave to take care of them. Jahoodi was scarred and battered. Tavia managed to get away without breaking a nail. A1 lost his entire crew and was out for revenge. As soon as the coast was clear, Tavia initiated the shockwave and then they were back in control of the FoF. Tavia made sure they had removed all the control codes and then they headed straight to activate the seismic reactor.

Frequency and Teez were buzzing around in the breacher. They got close to drop area, where Frequency had given the crew to drop the container. Frequency contacted crew and told them to run the Foundation protocol. Everybody went to their locations and followed the instructions accordingly. CMAX

drew up a large amount of power blacking out half of the city, all the robots joined within the tank and became liquified. They all then were injected within the underwater container that took the shape of the vessel Teez named Lotus Flower.

Lotus Flower

The entire crew loved the new hangout. It had enough room for all the sea breachers and the Airfish. Frequency figured if he was in the Lotus Flower no one could sneak up on him again. Teez was amazed at how Frequency took all her design cues and made this a lot about her. Teez enjoyed knowing she had someone that was there to take care of her no matter what. Thunderhead was sitting up in the Captain's chair when he saw something that he hoped to never see, it was the FoF.

The alerts went off in the ship and everybody ran to the bridge.

Teez couldn't believe it, somehow that crazy bastard was able to get free and he was coming after them. Frequency told everyone to go to the main room and allow him and the crew deal with this maniac once and for all. Everybody ran and got seated. Frequency then activated all the monitors so that everybody could see what was going on. Teez fired up the Lotus Blossom and headed straight for the FoF. Frequency then told everybody that what was so special about the ship, that it was not a ship, but a submarine. As he said that the ship begins to dive deep and go under the FoF radar system untraceable. Teez pointed the sub to the location of the seismic reactor. If the FoF was back in the hands of Jahoodi, then that reactor would be the next item on his list.

Jahoodi was livid, he had lost a fucking sub to Frequency Hunter and he wanted everybody to suffer. He screamed at Tavia to explode the reactor as soon as possible. Only problem was they had gone too far out of range to detonate it remotely. Tavia then told him Frequency was headed straight for it. L1 knew they needed to get that island out to sea to so that Jahoodi could continue to be a free man. L1 told Jahoodi to get in the helicopter and let's fly out close enough to detonate the reactor, meanwhile Tavia can get the FoF out in the ocean where we can regroup and recoup our loses. Jahoodi knew L1 was right but he didn't like to be told what to do, he decide to go with L1 ideal.

    Frequency and the crew were loving the underwater view they made it to the reactor and was starting to cut the cable that had it anchored to the ocean floor. The radar picked up a helicopter flying low then something hit the water and rocked the Lotus Blossom. Jahoodi dropped a human torpedo into the water and it worked like a charm. Jahoodi liked L1 but he didn't like being to what to do by anyone.

Frequency awoke first and immediately ran to the bridge of the sub. Everything was offline and to make matters worse the cable was caught into the propellers. Frequency immediately grabbed a rebreather and wet suit. He got into the dive chamber and exited the sub on a solo repair mission. Teez and the rest of the crew slowly awoke then assessed the situation. By the time they had figured out what had happened, it was too late to do anything about it. Frequency was working fast getting the cable out of the props. He swam around to one of the cameras and gave a thumbs up. Teez tried to power up the sub but nothing happened. Frequency saw what the problem was, and he was headed to go fix it when the cable snapped from the ocean floor. It hooked a piece of his dive gear as it ascended from the depths. Suddenly the reactor exploded but instead of the shockwave killing everybody in the area, it knocked Frequency into the sub and his nanocamo skin engulfed the entire Lotus Flower. Teez and the whole crew were knocked around, but they were unharmed. Amazingly, the sub reacted on its own. Teez was confused as all the automated system were now online. Teez immediately thought about Frequency, he was nowhere to be found.

Jahoodi had landed back on the FoF and got in touch with his ground forces. They told him that everything was looking bad and that Mr. White had escaped once again. Jahoodi main dude stopped talking when a bullet hit him. Ms. Bobbi got on the video and let Jahoodi have an earful, about how he can't kill a real queen. She then showed him that the GayMack and her man Mario was still standing tall. With that note she told Jahoodi to kiss her ass, balls and all. Jahoodi was beyond pissed he looked and saw that Hunter had left one of the seas breachers. Jahoodi called one of his mechanics and asked him if he could fix this for him but add a little extra. The mechanic had it ready in no time and it ran on one of Tavia's mini reactor.

Tavia decided to go with Jahoodi and show him how to kill without remorse. The sea breacher was silent as it sliced through the dark water. It wasn't long before the two of them was climbing out of the breacher and back into Jahoodi's Genesis. Jahoodi went to his hideout and saw that Ms. Bobbi had did a number in there. Jahoodi waited and his assassin crew arrived. They gave him the information he needed and he in turned shot them all. Tavia set the detonator on the C4 and the two left. Next, they went to a hotel got cleaned up and waited for the next explosion.

    The Reverend lay in the jail hospital bed recovering from his wounds. He and Madame lived to see another day. He knew he could recover from this scandal because his flock believed in him. He and the others just had to get to a courthouse and then it would all be cleared up. The jailhouse shook as the courthouse collapsed from the explosion. With that Jahoodi had set off a chain reaction of events, that also included activating the viruses. Every man, woman, and child would suffer from the TFV or PFV by daylight.

The hospitals were full, the people were coming into the ER by the hundreds. Men were stuck inside of the lovers because of the Blowfish Virus or they were becoming eunuch from the Toothpick Fish Virus. The women were no worse for the wear. It would be their tongue or hand with the Blowfish Virus or eunuch with the ToothPickFishVirus. The hospitals were in full desperation mode and the CDC had no answers. Jahoodi had Tavia send out the recording that would put him on the death strike list. The recording just said "Infidels" you will fuck no more!!!! Jahoodi and Tavia decided to leave the country and wait on the highest bidder for the cure.

    Ms. Bobbi lay in the floor screaming in pain, her genitals looked like a puffer fish. She could barely walk and wondered why GayMack was not in pain. The GayMack laughed because his dick sock saved him. He had been wearing it ever since he got up out of the wheelchair. He decided to make it a fashion assessory. He looked and saw Ms. Bobbi and he called Dr. Stamps, who was having problems of his own. GayMack went into the room and found Dr. Stamps stuck in Dime's corpse. Dime didn't survive the water. He also found the corpses of Quarter and Nickel. They succomed to the virus while in a 69 position. GayMack asked Dr. Stamps could he fix this or is this how they will find you dick stuck in a corpse.

Mr. Mario was scared he made it back to the city but now everybody was getting sick. Mario was about to get on a bus, when he heard some kid tell his uncle that he was the dude that raped him. Mario looked up and saw PPK running straight for him. Mario immediately got to running franticly through the bus station, he saw a basement building grate open, so he jumped inside it. The grate closed over the top of him, but he could hear PPK crew talking and looking for him. Mario got comfortable because he wasn't leaving until the coast was clear. The pain that Mario felt was agonizing the building grate he decided to lay in was PPK's dry cleaners. They found him in there and decided to let the women iron his genitals before they even begin to remove them. Mario would not die fast, and he would not see another day.

Jahoodi sat in the car as Tavia set the last of the explosions she got into the car and the two headed for the dock. The drive was pleasant when suddenly, the Genesis was sent flying.

BDB stopped the truck call Muthafucka and let Teez out. Teez went over to the passenger side of the vehicle and helped Tavia out. Teez asked her was she going fight or run, Tavia took off into the grassy field. Teez looked at BDB and he told her to handle your business. Bossman wasn't going anywhere.

Jahoodi was totally shook, he knew he had no chance with this behemoth of a man. BDB pulled out a testosteroid shot and went to the car and administered it to Jahoodi. Jahoodi got out of the wreckage and immediately started negotiating with BDB.

Jahoodi offered the big man millions of dollars to let him go. Jahoodi popped the trunk and there was enough money to make the entire crew millionaires. Jahoodi grabbed the case with the antidote, but he forgot one thing, he had no wheels and his woman was probably getting her ass kicked. Jahoodi reached for his weapon and drew it. The truck was there but the big man was gone. Jahoodi looked around then he went to go help Tavia. Teez was surprised, Tavia decided to stop running and start fighting. She under estimated her and now they were really throwing down. Tavia had nothing to lose so she just started fighting back and she caught Teez off guard. Tavia was on top slamming her head into the ground but she bucked her off and returned the favor. Tavia started grabbing for anything to hit her with and they both just had nothing but their hands and heart. Teez remembered what Frequency used to say and she dropped her full weight on Tavia knocking the wind from her. She then threw down hard with her elbow, catching her dead on the chin knocking her unconscious. Teez was about to hit her again when a shot rang out. It was Jahoodi and he didn't look happy. Jahoodi told Teez to get off his woman and step over to the side.

    Jahoodi asked Teez why you would attack me knowing that I would kill you. Teez told Jahoodi that he killed Frequency and now she was going to return the favor. Jahoodi reminded her that he had the gun as he pointed it at her, but then he started to choke. Jahoodi dropped the gun and then staggered back. The gun seems to rise in thin air and a familiar voice spoke. Frequency stepped from out of the nanocamo and Teez felled to her knees, he was alive. Jahoodi looked and tried to rush Frequency and the two men went into battle. BDB came down to the area and made sure Teez was ok.

Teez ask BDB did he get her purple pouch from the truck. BDB handed it to her and she went over to finish Tavia. Tavia had gotten up and made it over to the gun. Teez dropped her pouch as jumped to prevent being shot by Tavia. BDB moved fast, he grabbed the nanocamo and covered himself and Teez with it. Tavia immediately grabbed the pouch, then ran back to the main road where she hijacked some dude for his 4-wheel drive truck. Tavia immediately went looking for Jahoodi.

Jahoodi martial arts training paid off, he was able to dish out a beating as well as take one. Frequency was holding his own just waiting on this guy to make a mistake, so he could make him pay for it. Jahoodi then kicked Frequency in both knees, the groin area, and then a knee to the chin knocking him into a tree. Jahoodi tried to punch Frequency but he moved in time to catch Jahoodi with a downward punch to his left knee then back up with an elbow directly to his chin.

Jahoodi grunted and fell back hard on his back. Frequency the stomped down hard directly into Jahoodi groin doubling him up in pain. Frequency gathered his composure and then he soccer kicked Jahoodi in the ribs, when a shot just barely missed his ear. Frequency ran for cover as Tavia showed up to save Jahoodi again.

Tavia helped Jahoodi get into the truck and they hurriedly made it to the breacher and headed back to the FoF. Tavia went through Teez's pouch and it had moisturizer and lip gloss. She used it as her lips had become chapped. After Jahoodi and Tavia made it back to the FoF, they kissed and said live to fight another day.

Unknowingly to Tavia the lip gloss had the sea urchin venom in it. Their fighting days was numbered.

It seemed like it was just yesterday that Teez was sitting in a clinic just helping people. Now she was living on the Lotus Flower trying to keep this crazy crew out of trouble. They found the antidote for the viruses and a cure was being used. Tavia and Jahoodi were found dead on the FoF after they had accidently got ahold of a sea urchin. The rest of the crew either died from the virus, living the life eunuch, or still on the run. As for Frequency Hunter, he is just living in the life of the FREAKS.

# The End

www.ingramcontent.com/pod-product-compliance
Lightning Source LLC
Chambersburg PA
CBHW062020290426
44108CB00024B/2724